Conference Proceedings

RTIE 2016

UGC Sponsored

Second National Conference

On

Recent Trends in
Instrumentation & Electronics

(RTIE 2016)

5th and 6th October 2016

Compiled and Edited by:

Dr Amita Kapoor
Secretary, RTIE 2016

Dr Sneha Kabra
Joint Secretary, RTIE 2016

Special thanks to

Prof Enakshi K Sharma
Member Advisory Committee,
Head, Department of Electronic Science,
University of Delhi South Campus

Prof Mridula Gupta
Technical Program Chair
Department of Electronic Science,
University of Delhi South Campus

Venue: Shaheed Rajguru College of Applied Sciences for Women,

Vasundhra Enclave, Delhi 110096, India.

ISBN-13: 978-1539305422
ISBN-10: 1539305422

Chairperson
Dr Payal Mago, Principal, SRCASW
Secretary
Dr Amita Kapoor, Associate Professor, Department of Electronics, SRCASW
Joint Secretary
Dr Daya Bhardwaj, Assistant Professor, Department of Instrumentation, SRCASW
Dr Sneha Kabra, Assistant Professor, Department of Instrumentation, SRCASW
Treasurer
Dr Punita Saxena, Associate Professor, Department of Mathematics, SRCASW
Dr Jasjeet Kaur, Associate Professor, Department of Chemistry, SRCASW

Members Advisory Committee

Dr. Amod Kumar, Acting Director, CSIR.
Prof Anurag Sharma, Department of Physics, IIT Delhi.
Ms Anuradha Sharma, Social Innovation Program Manager, Cisco Systems, India.
Ms Archana Jyoti, Pioneer Group.
Prof. Avinashi Kapoor, Department of Electronic Science, UDSC.
Prof. A. K. Verma, Retd. Prof, UDSC.
Prof. Enakshi Khular Sharma, Head-Department of Electronic Science,UDSC.
Prof Dr. Hukum Singh, Former Dean Academic and Head DESM and DEK, NCERT, Delhi.
Dr Jasdeep Kaur, IGTDU
Mr Lokesh Mehra, Director Education services,Symantec, India.
Prof N. Raja, School of Technology and Computer Science, TIFR, Mumbai, India.
Prof. Pramod Kumar Bhatnagar, BSR Fellow, Department of Electronic Science, UDSC.
Dr. Rahul Singh, Head – CQA (HCD) & R&D (Analytical), Emami Ltd.
Prof. R. S. Gupta, Head – Department of Electronic & Communication Engg., M.A.I.T, GGSIPU.
Prof R. M. Mehra, Sharda University.
Prof. Santanu Chaudhury, Director, CSIR, Pilani, India.
Prof. Sneh Anand, IIT Delhi.
Prof. S. Saiful Islam, Hon. Director, Center for Nanoscience and Nanotechnology, JMI.
Prof. S. S. Agrawal, Executive Director and Prof. ECE and CSE, KIIT Group of Colleges.
Prof. Vijay K. Arora, Department of Electrical Engineering and Physics, Wilkes University, USA.
Prof Wolfgang Freude, Karlsruhe Institute of Technology, Germany.
Dr Yogesh Valenkar, IUCEE, Centre of Excellence.

Members Organizing Committee

Dr Alka Vohra Kuan, Associate Professor, Department of Physics, SRCASW.
Dr Amit Pundir, Associate Professor, Department of Electronics, MAC.
Dr Amit Garg, Associate Professor, Department of Electronics, ANDC.
Dr Amita Kapoor, Associate Professor,Department of Electronics, SRCASW
Dr Daya Bhardwaj, Assistant Professor, Department of Instrumentation, SRCASW
Dr Geetika Jain, Associate Professor, Department of Electronics, MAC.
Dr G S Sodhi, Associate Professor, Department of Chemistry, SGTB Khalsa College.
Dr Jasjeet Kaur, Associate Professor, Department of Chemistry,SRCASW
Dr Manoj Saxena, Associate Professor, Department of Electronics, DDUC.
Dr Punita Saxena, Associate Professor, Department of Mathematics,SRCASW
Dr Saakshi Dhanekar, INSPIRE Faculty Fellow, IIT Delhi.
Dr S. Haldar, Associate Professor,Department of Electronics, DU.
Dr Sneha Kabra, Assistant Professor,Department of Instrumentation,SRCASW
Ms Venika Gupta, Associate Professor, Department of Electronics, SRCASW.

Foreword by the Principal

It gives me an immense pleasure that our college is organizing a two-day UGC-sponsored second **National Conference on Recent Trends in Electronics and Instrumentation**, RTIE 2016 to be held on 5th and 6th October.

Research and innovation are the backbone for the success and progress of any nation. The teaching faculty of various colleges, universities, technical institutes and professionals working in industry and research scholars from various organizations need to join hands for the overall development of the society. Such conferences provide an ideal platform to share their research endeavors and to provide direction for future research in the specific area. The conference covers the relevant topics in the field of Instrumentation and Electronics. Its technical program includes invited talks, oral presentations, poster presentations and panel discussions on recent advancements in these disciplines. A special session on Energy Efficient Systems is being organized to cater to the need of the hour.

I extend my gratitude to UGC and DBT Star College Scheme for sponsoring the conference. Thanks are due to IEEE EDS, Delhi Chapter and TipzIn for extending their support. We are honored to have **Sh. Manish Sisodia**, Hon'able Deputy Chief Minister, Govt. of NCT of Delhi, as the Chief Guest and **Prof. M. Jagadesh Kumar**, Vice Chancellor, Jawaharlal Nehru University as Guest of Honor for the Inaugural Function. I am thankful to **Prof. Mridula Gupta**, Chairperson, IEEE EDS Delhi Chapter, Department of Electronic Science, University of Delhi; **Prof. Enakshi K. Sharma**, Head, Department of Electronic Science, University of Delhi; **Prof. Avinashi Kapoor**, Department of Electronic Science, University of Delhi for gracing the occasion.

I would like to congratulate the organizing committee and advisory committee members for making good arrangements for the smooth conduct of the conference and wish them all success. I also hope that the event will prove to be a great learning experience for all the delegates.

Dr Payal Mago
Chairperson, RTIE 2016
Principal,
Shaheed Rajguru College of Applied Sciences for Women,
University of Delhi,
Vasundhra Enclave, Delhi.

Foreword from Secretary

I welcome you to the *Second UGC and IEEE EDS sponsored National Conference on Recent Trends in Instrumentation and Electronics* (RTIE) held October 5–6, 2016 in the Shaheed Rajguru College of Applied Sciences for Women. As a premier conference in the field, RTIE 2016 provides a highly competitive forum for reporting the latest developments in the research and application of Instrumentation and Electronics. I am pleased to present the proceedings of the conference as its published record.

RTIE is a young conference for research in the areas of electronics, instrumentation and related fields. This year the theme had been energy efficient electronic systems. The conference has five sessions with the themes namely: Information and Communication Technology & Internet of Things, Industrial Automation & VLSI, Energy Efficient Systems, Analytical Instrumentation, Sensors and Actuators, BIOMEMS and Data Encryption techniques. Although it is only in its second year, it has already witnessed significant growth.

The conference program represents the efforts of many people. I want to express my gratitude to the members of the Advisory Committee and Organizing Committee, **Prof Enakshi K Sharma,** Head Department of Electronic Science, Technical Program Chair **Prof Mridula Gupta,** and the external reviewers for their hard work in reviewing submissions. I am thankful to the Chief Guest **Sh Manish Sisodia** (Honorable Deputy Chief Minister, Govt. of NCT of Delhi), our Guest of Honor **Prof. M Jagadesh Kumar,** (Vice-Chancellor, Jawaharlal Nehru University) for sparing their valuable time with us. I am also thankful to our plenary speakers, **Dr. Amod Kumar,** (Chief Scientist, Business initiatives and Project Planning, CSIO-Central Scientific Instruments Organization, Chandigarh.) and **Prof Deepak Garg** (Head, Computer Science Engineering Department, Bennett University, Greater Noida), for sharing their insights with us. The chairperson of the conference **Dr Payal Mago** (Principal SRCASW), have helped us in many ways, for which we are grateful. Finally, the conference would not be possible without the excellent papers contributed by authors. I thank all the authors for their contributions and their participation in RTIE 2016!

I hope that this program will further stimulate research in the fields of electronics and Instrumentation, and provide practitioners with better techniques, algorithms, and tools for deployment. I feel honored and privileged to serve the best recent developments in the field of instrumentation and Electronics to you through this exciting program.

Dr Amita Kapoor
RTIE 2016, Secretary
Associate Professor,
Department of Electronics
Shaheed Rajguru College of Applied Sciences for Women,
University of Delhi,
Vasundhra Enclave, Delhi

Foreword From Joint Secretary

The seeds of this conference were sown about a year and half ago when Department of Instrumentation and Electronics of the college hosted the first National conference, "Recent trends in Instrumentation and Electronics" in January 2015. It was only after the grand success of the event that we decided to organize the second edition of the conference, RTIE 2016.

The aim of the event is to provide a platform to researchers, academicians and people from industry to share their ideas and enhance the knowledge. The programme sessions have been carefully designed so that the latest advancements in the field of Instrumentation and electronics such as Engineering Optimization, Internet of Everything (IoE), Information and Communication Technologies (ICT) and Instrumentation and Measurements can be discussed and analyzed. Along with the topics mentioned above, special session on Energy Efficient Systems was also conducted in the RTIE 2016.

The conference program represents the efforts of many people and support by financial bodies. We would like to convey special thanks to our sponsors **UGC, IEEE EDS** Delhi Chapter, DBT Star college scheme and TipzIn who have provided us financial support to organize this event. We are thankful to **Prof. Mridula Gupta**, technical chair of the conference for providing technical support for the conference; **Prof. Enakshi K Sharma**, Head, Department of Electronic Science, University of Delhi South Campus for her guidance and **Dr. Payal Mago**, Principal, Shaheed Rajguru College of Applied Sciences for Women for her unconditional support. We sincerely thank all the invited speakers for sharing their insights with us. We would also like to appreciate the tireless efforts put in by the core committee and organizing committee members of RTIE 2016, without whom, organizing this event would not have been possible.

We wish all the participants good luck and happy learning throughout the two days of the conference.

Dr Daya Bhardwaj,
RTIE 2016, Joint Secretary
Department of Instrumentation
Shaheed Rajguru College of Applied Sciences for Women,
University of Delhi,
Vasundhra Enclave, Delhi

Dr Sneha Kabra,
RTIE 2016, Joint Secretary
Department of Instrumentation
Shaheed Rajguru College of Applied Sciences for Women,
University of Delhi,
Vasundhra Enclave, Delhi

Index

An Automatic Irrigation System Using Self-Made Soil Moisture Sensors and Android App.

Shalu Sharma
Department of Mathematics
SRCASW, University of Delhi
Delhi India

Shivani Seth, Tanya Gandhi
Department of Instrumentation
SRCASW, University of Delhi
Delhi, India

Suruchi Chawla, Shakshi Bachhtey, Veni Gupta, Bhanvi Shukla, Monica Gupta, Pragya Kaushik, Shreshtha Pushkar
Department of Computer Science
SRCASW, University of Delhi
Delhi, India.

Sheetal Varshney, Saloni Mehta, Ruchika Jha, Amita Kapoor
Department of Electronics
SRCASW, University of Delhi,
Delhi, India

Abstract—**This paper presents the design of an automatic irrigation system using low cost soil moisture sensor and android App. We develop an app based automatic irrigation system using self-made capacitive sensors. Then sensors are interfaced with Raspberry Pi microcontroller. To detect the threshold levels of moisture and control the inflow of water for optimal use of water an algorithm is developed. We have developed an android app using Java to Interface Raspberry Pi microcontroller. Thus, we have an automatic irrigation system, which can be used to increase the productivity of crop by providing optimal amount of water.**

Keywords— automatic irrigation; self-made sensors; irrigation system; Microcontroller Android Application.

I. INTRODUCTION

Drip irrigation for efficient irrigation is being used by developed countries farmers. In this system water is supplied near to the root zone of the plants drip by drip, thus ample amount of water is saved and plants get adequate water simultaneously. In India, farmers manually irrigate the fields at regular intervals. This process consumes more water and sometimes crop does not get water at right place, which can decrease the productivity of crop. Deficiency of Water can be risky to plants before noticeable wilting occurs. This problem can be resolved if we have a drip irrigation system in which the irrigation will happen only when the plants will have requirement of water.

Due to the limited resources of water, there has been an interest in the researchers for developing an automatic irrigation system. Various irrigation system like use of canopy temperature [1], array of wireless sensors [2] [7] [8] [10], regulating soil water tension with on/off strategies [3],[4] based on the feedback, android based [11] [12], small embedded system device [9] , Raspberry pi based system [13] and using controllable parameter such as Temperature, soil moisture and air humidity [5] have been proposed by the researchers. These systems are quiet useful but are money extensive, and hence not suitable for Indian farmers.

In this paper we present an app based automatic irrigation system using a self-made, low cost sensor and Raspberry microcontroller. The system sets the irrigation time based on the moisture reading from the sensors and irrigates the field automatically, when unattended. We have used a capacitive sensor to measure soil moisture, which is inserted at various positions near to the plants. The Raspberry microcontroller receives the sensor output. We take the level of water sensed by sensors as input to decide the manual/automatic on or off of motor controlling the supply of water to plants. This system is useful for people who don't have much time for watering their plants. For automatic irrigation we use an android app based on sensor input and microcontroller. The App retrieves the values from database, which is stored in Raspberry Pi that contains data sent from the sensors.

II. SELF-MADE SENSOR

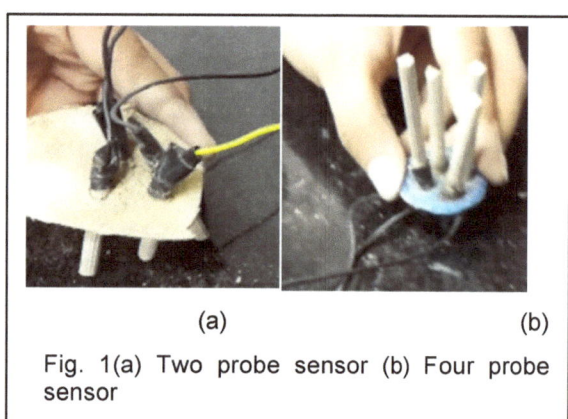

(a)   (b)

Fig. 1(a) Two probe sensor (b) Four probe sensor

In 1995 Gluck [5], proposed dielectric soil moisture sensors, he used the change in dielectric property of the soil with water as a parameter. In our sensors we take two/four galvanized aluminum rods (Fig 1). We buried these rods deep into the soil near the roots of the plant. The resistance between the rods gets modified, when moisture level of the soil changes.

 We used galvanized aluminum rods because they are inert and they are easily available. AC source has been used in most soil sensor circuits as they measure the capacitive properties of soil. For AC source we need a special circuit, which makes it not convenient for use by local Indian farmers. Thus, we use dc power source in our soil sensor.

DC ANALYSIS OF THE SENSOR

Two rod sensor can be approximated as a RC circuit in parallel. In DC Analysis, soil resistance dominates the circuit and the capacitance comes into picture only in the absence of moisture.

Under DC biasing conditions, the drop across R1 (this corresponds to the resistance offered by soil) is given by voltage divider. The resistivity of the soil

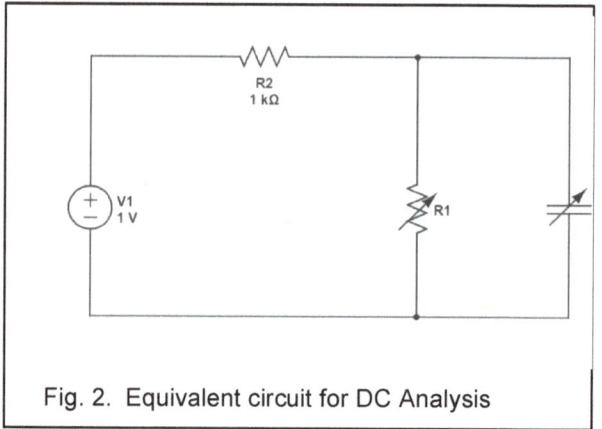

Fig. 2. Equivalent circuit for DC Analysis

depends on many factors like its salinity, temperature, moisture level etc. It is possible to choose the distance between the two aluminum rods such that the resistance depends inversely on the moisture content of soil, and voltage drop across rods is given by:

$$V_{R1} = \frac{\dfrac{\alpha}{R}}{\dfrac{\alpha}{R} + M} V$$

Our test results demonstrated that for such a linear dependence on 1/M the distance between the probes should be kept between 2.5 cm-5 cm.

III. RESULTS

We first measured the variation in the change in voltage due to the presence of moisture as a function of distance between the probes. If V_D is the voltage across probes when soil is completely dry, and V_W is the voltage across the probes when soil is supplied with 100ml of water, then $\Delta V = V_D - V_W$ Fig 3 shows the plot between Δ and distance between the probes. It can be seen from the graph that between 2.5 cm-5 cm, the relationship is linear.

Then, we investigated how the voltage across probes' vary with the soil moisture content. we found that as the moisture level increases, the voltage across the probe decreases as accepted, also the voltage variation is linear for 3.5 cm and 4 cm distances probes.

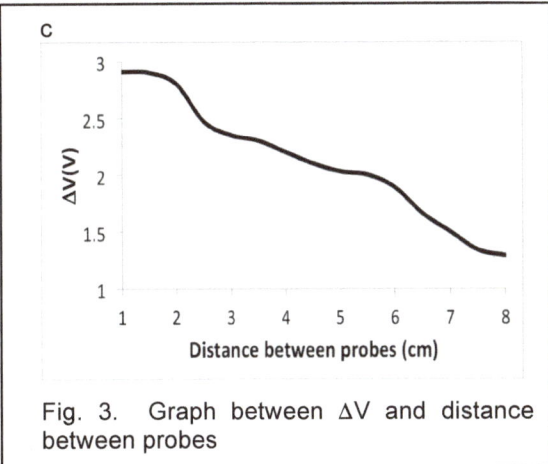

Fig. 3. Graph between ΔV and distance between probes

The output of the sensor is then fed to an differential amplifier (LM393), at present we are using only threshold detection, that is the amplifiers gives a high voltage when soil is in dry condition (below certain threshold moisture) and generates low signal once the threshold moisture is reached. This digital signal is then fed to a microcontroller (Raspberry Pi II), which then controls the switching on and off of irrigation system.

ANDROID APP

We develop the Android app using Java to interface with raspberry pie microcontroller. The app provides the user interface to verify the user identity for automation of irrigation system. To authenticate the identity of authenticated users, the user login information is stored in the database and is retrieved. After doing successful login the sensor data request can be sent to microcontroller and to identify the water level in plants the requested data can be retrieved from the sensor database. Thus upon receiving the sensor data on android app it is processed using threshold level of water in order to give on/off command to microcontroller. Hence the microcontroller upon receiving the on/off command from user, the motor is on/off accordingly for the supply of water to plants.

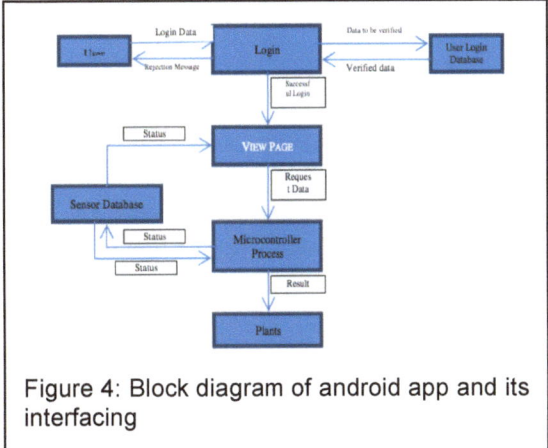

Figure 4: Block diagram of android app and its interfacing

We use python script on microcontroller to process the command received from android and store the sensor data in database. The block diagram of the android app and its interfacing is given above in Figure 4

The proposed system is given below in Figure 5. Wifi/3G Network has been used to interface the android app with the server. The interfacing of

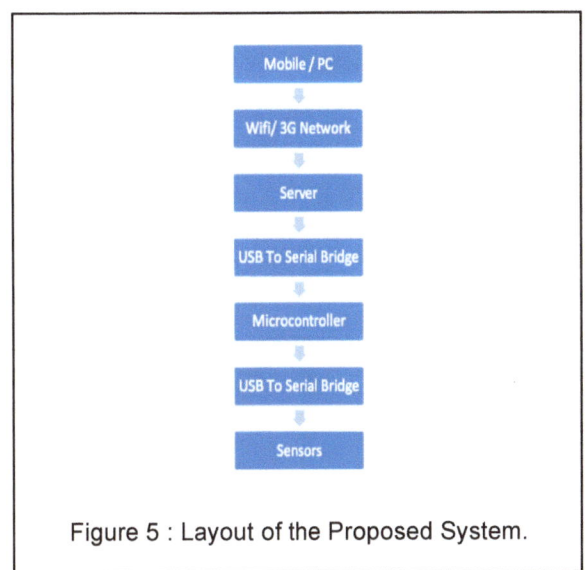

Figure 5 : Layout of the Proposed System.

microcontroller with server and sensor is done using USB to serial bridge.

The following steps are involved in the working of the system:

1. Start
2. Database created on the Raspberry Pi's Apache server

3. Sensor readings are stored in database
4. App retrieves water level reading of soil
5. App commands the action to be performed
6. Server sends the signal to the system.
7. End.

The experiment was conducted to test the automatic control of irrigation using android-based app, Raspberry Pi and low cost capacitor sensors. The hardware and software required for the experiment are an android phone/emulator from android studio, Raspberry Pi (2B), Internet connection with port 80 forwarded to the IP of Raspberry Pi. Android studio, Raspbian OS for the Raspberry PI, Apache 2 for the server to be run on the Pi, PHP to create, modify and manage the databases and MySQL for the databases.

The database is created using server side php scripting. On client side java , json parser library is used for making the app and the server communication. JSON Parser library is used for parsing the PHP code to Java and vice-versa, th us making app and the server, both, to comprehend the parameters correctly and the Http connection is established. The Snap shot of json library and http connection with Raspberry is given below in Figure 6.

The Android studio emulator is used for simulation of app communication with raspberry pie for send/receiving data based on post method. This is given in fig 7.

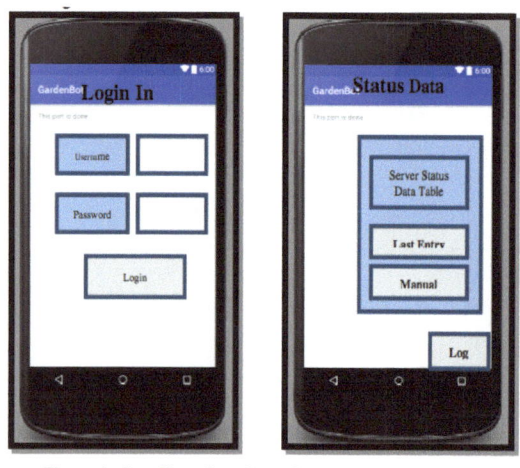

Fig 7: SnapShot of emulator simulation of Client App and its GUI interface.

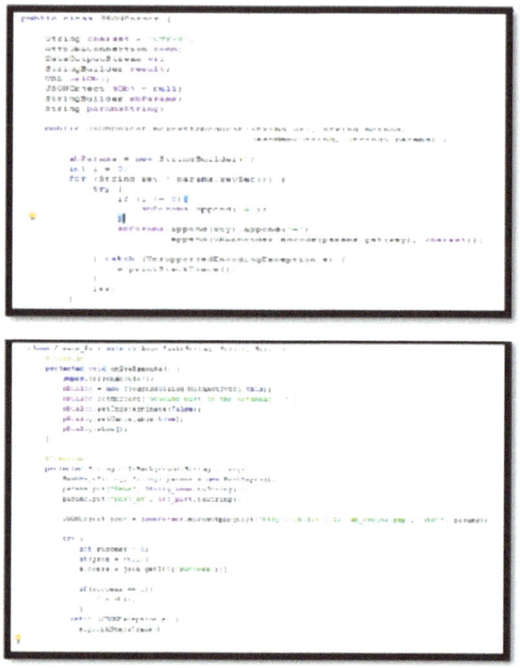

Fig 6. SnapShot of JSON Parser and http connection with Raspberry Pi

IV. CONCLUSION

An android-based automatic irrigation system is designed using a self-made capacitor sensor with galvanized aluminum rods and interfaced with Raspberry Pi microcontroller. The android app is encoded to retrieve the sensor data from Raspberry Pi microcontroller and enables the user to supply the water to field automatically. The design is low cost, low power, robust, small size and highly versatile. Thus, this system avoids over use of water in irrigation, under irrigation and reduces the wastage of water. The present work is an attempt to save the resources for human kind which are available naturally, and provide Indian farmers a low cost automatic system for irrigation. we can control the flow of water by constantly observing the status of the soil moisture, and then reduce the wastage of water. The main benefit is that the system can be modified according to the situation (plants, weather conditions, soil etc.).

ACKNOWLEDGMENT

The authors are thankful to University of Delhi India for financially supporting the project under "Innovation Project Scheme". Thanks are also due to our mentor, Prof Enakshi K Sharma, Head, Department of Electronics, University of Delhi South Campus for stimulating dialogues, constant guidance and unconditional support

REFERENCES

[1] Evett, Steven R., et al. "Canopy temperature based automatic irrigation control." Proc. Intl. Conf. Evapotranspiration and Irrigation Scheduling. 1996

[2] Arun, C., and K. Lakshmi Sudha. "Agricultural Management using Wireless Sensor Networks-A Survey." 2nd International Conference on Environment Science and Biotechnology (IPCBEE), Singapore. 2012.

[3] Luthra, S. K., et al. "Design and development of an auto irrigation system." Agricultural Water Management 33.2 (1997): 169-181.

[4] Boutraa, Tahar, et al. "Evaluation of the effectiveness of an automated irrigation system using wheat crops." Agriculture and Biology Journal of North America 2.1 (2011): 80-88.

[5] Gluck, Israel, Anatoly Friedman, and Naftali Feniger. "Soil moisture sensor" U.S. Patent No. 5,424,649. 13 Jun. 1995.

[6] Al-Ali, A. R., Qasaimeh, M., Al-Mardini, M., Radder, S., & Zualkernan, I. A. (2015). Zig Bee-based irrigation system for home gardens. In International Conference on Communications, Signal Processing, and their Applications (ICCSPA), 1-5, IEEE.

[7] Gutiérrez, J., Villa-Medina, J. F., Nieto-Garibay, A., & Porta-Gándara, M.Á. (2014). Automated irrigation system using a wireless sensor network and GPRS module. IEEE transactions on instrumentation and measurement,63(1), 166-176.

[8] Jadhav,S., & Hambarde, S.(2016). Android based Automated Irrigation System using Raspberry Pi, International Journal of Science and Research, 5(6),2345-51.

[9] Malage, S & Bhole, K (2015). Low cost remotely operated smart irrigation system, In International conference on Industrial Instrumentation and control (ICIC),1501-1505, IEEE.

[10] Nisha, G., & Megala, J. (2014). Wireless sensor Network based automated irrigation and crop field monitoring system. In 2014 Sixth International Conference on Advanced Computing (ICoAC) ,189-194,. IEEE.

[11] Panth, S& Jivani,M(2013). Home automation system using android for mobile phone. International Journal of Electronics and computer Science Engineering ISSN-2277-1956.

[12] .Dursun, M., & Ozden, S. (2011). A wireless application of drip irrigation automation supported by soil moisture sensors. Scientific Research and Essays, 6(7), 1573-1582.

[13] Chate, B. K., & Rana, J. G. (2016). SMART IRRIGATION SYSTEM USING RASPBERRY PIComputation of Bandwidth Utilization in a Network File System

Computation of Bandwidth Utilization in a Network File System

Tina Sachdeva,
Assistant Professor,
Department of Computer Science
Shaheed Rajguru College of Applied Sciences for Women,
University of Delhi,
e-mail:sachdeva_tina@yahoo.com

Nehal Sharma,
B.Tech Computer Science (4th year)
Shaheed Rajguru College of Applied Sciences for Women,
University of Delhi,
Email id: nehal.srcasw.du@gmail.com

Abstract - **There is a huge load of data on the network, which needs to be accessed by a number of systems simultaneously, and placing it on all systems leads to inefficient use of disk space. Further, it also leads to data redundancy and inconsistency. This problem can be solved by a Network File System (NFS) which allows remote hosts to mount file systems over a network and interact with those as though they are present locally. The aim of this research paper is to compute bandwidth utilization in NFS. This paper also describes the method of improving bandwidth utilization by exploiting the common content among different versions of the same file. The effect of this improvement on the bandwidth has also been discussed in the paper.**

Keywords: Bandwidth utilization, Computer Networks, Data Transfer, Network File Systems.

I. INTRODUCTION

Sun Microsystems designed the Network File System (NFS), a client-server application that manages files on multiple computers in a network as if they were present on their local hard disks. An NFS client mounts a remote file system onto its local file system namespace and makes it behave like a local UNIX file system. Multiple clients can mount the same remote file system so that users can share files for ease of collaboration. NFS has been developed by Rick Sladkey, who is also credited with the development of NFS kernel code and large parts of the NFS server.

ADVANTAGES OF NFS:
NFS offers a number of advantages:

1. Utilization of disk space at local workstations is reduced since the most frequently used information is stored at a single place and is still accessible.
2. Home directories could be set up on the NFS server and made available throughout the network.
3. Portable storage devices are not required.
4. Data consuming large amounts of space and administrative data may be kept on a single host.

WORKING OF NFS

NFS works in conjunction with the Virtual File System (VFS).The NFS resides at the client and it sends NFS requests in the form of RPCs (Remote Procedure Calls) to the server. The job of NFS is to convert requests like read() and write() into RPCs and then sends them to the server. The NFS resides in the *user space* at the server side and takes RPCs client's send requests and convert them into system calls that read and write files on the server's storage disk.

It is the server program's job to make the file systems accessible to other machines via a process called *exporting*. Such file systems are often called as shared file systems.

In order to access shared file systems, NFS clients mount them from an NFS server machine and then unify them into their directory tree. An *automounter* is a more progressive form of mounting that mounts and unmounts file systems automatically.

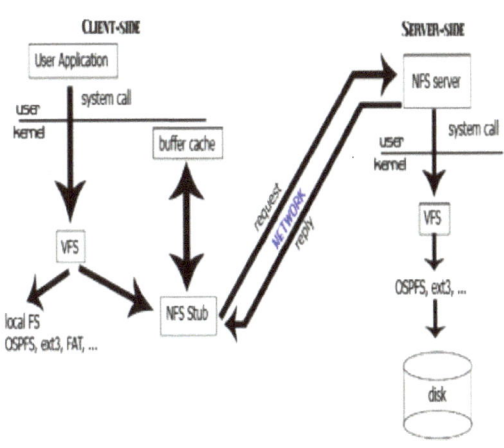

Fig. 1: *NFS client and server and their relationship to the VFS layer and kernel.*

[Courtesy: *Illustrations by Subhash Arja, Matthew Ho, Grant Jenks, Samuel Kwok*]

MEASURING NFS BANDWIDTH

SETTING UP THE STAGE

NFS is quite mature and widely used and a lot of support is also available. Thus, NFS has been critically scrutinized from a number of aspects but the main parameter has always been its bandwidth utilization. Lot of work can be related i.e., combined with the previous one to get additional savings.

Hardware Requirements:

A system serving as the production server with a large amount of disk space

A number of other systems acting as clients

Software Requirements:

1. Linux Operating System installed on all the systems.
2. Kernel version of Linux should be at least 2.4.2.
3. Mount version of the Linux kernel should be at least 2.6

4. NFS utilities compatible with the Linux kernel version and mount version.

Configuration of the Server:

The three important configuration files we need to edit to set up an NFS server are as follows:

1. /etc/exports: It consists of all the entries of file volumes that are shared across the network.
2. /etc/hosts.allow and /etc/hosts.deny

These files indicate the client machines on the network that can utilize the services of the server machine. They control access to NFS by restricting connections to the daemons that provide NFS services.

The first daemon to restrict access to is the portmapper, which informs the requesting clients how to locate all the NFS services. Thus, limiting access to the portmapper prevents unauthorized clients from invading our system through NFS as they would not be able to find the NFS daemons.

GETTING THE SERVICES RUNNING

Imperative Requirements

First of all, the relevant packages namely kernel and an appropriate version of the *nfs-utils* needs to be installed. TCP/IP networking also needs to be operational. We can use telnet, FTP etc. In the newer version of Linux, NFS simply runs by rebooting the machine, and the startup scripts should detect our /etc/exports file. The need for discarding our configurations might arise if *nfsd* was already running when we edited our configuration files above.

Starting the Portmapper

NFS depends on *portmap* or *rpc.portmap* (the portmapper daemon). This daemon starts functioning in the boot scripts in the most recent Linux distributions, but it is important to make sure that this daemon is up and working correctly (This can be detected by tying ps aux | grep portmap).

The Daemons

Five daemons responsible for the proper functioning of the NFS server are: *rpc.nfsd*, which performs most of the work; *rpc.lockd* and *rpc.statd*, which take care of file locking; *rpc.mountd*, whose job is to handle the initial mount requests, and *rpc.rquotad*, which manages user file quotas on exported volumes.

If the five daemons are not included in the startup scripts of the distribution, they should be included and arranged to work in the following sequence:

Table 1: Table showing NFS status obtained by issuing rpcinfo-p command

rpc:portmap

rpc:mountd, rpc.nfsd

rpc.statd,

rpc.lockd(if necessary),rpc.rquotad

VERIFYING THAT NFS IS WORKING

The present status of the portmapper can be found using the command *rpcinfo -p*.

II. CARRYING OUT TESTING

MEASURING BANDWIDTH

We devise an algorithm using the *nfstrace* and *nfsstat* commands to measure the bandwidth.

nfsstat command

This command displays statistical information. Default switch is –*csnr* which does not reinitialize anything but displays all the statistical information.

/usr/sbin/nfsstat [-c] [-s] [-n] [-r] [-z] [-m]

nfstrace command

A record is produced each time a file is opened, giving a summary of what occurred. The output format consists of 7 fields:

timestamp |command-time | direction | file-id | client | transferred | size

III. RESULTS

FORMAT OF THE REPORTS THAT THE PROGRAM GENERATED

The major outputs of the program showed bandwidth utilization in the form of reports (statistical information about the number of calls received, number of bad calls, number of unauthenticated calls etc. and also timings taken for their execution). This output was converted into a graph, which gave a snapshot of the bandwidth.

FORMATS OF THE REPORTS

The client program provided the interface of the following type:

Name of the remote file to be accessed: _____

Path of the remote file to be accessed: _____

Sorry, the server for the requested file did not authenticate you!

Or

The server for the requested file has authenticated you! You may proceed further!

Menu

Open the file for reading

Open the file for writing

Open the file for execution

Exit

The server took _____ sec to authenticate your call and send you the desired file.

UTILIZATION OF BANDWIDTH

(output from the program)

The x-axis represent the size of the files (in kb) while the y-axis represents the time (in milli-seconds) taken to fetch it. We observe that the performance of NFS is not too high.

Table II Showing

File Size(kb)	Time(msec)
8.3	856.7
8.4	854.6
8.5	871.5
9.0	820
9.1	800
9.2	798.3
9.3	671.7
9.4	786.4
9.5	734.2
10.0	769.6

Consumption pattern of NFS over Linux client and

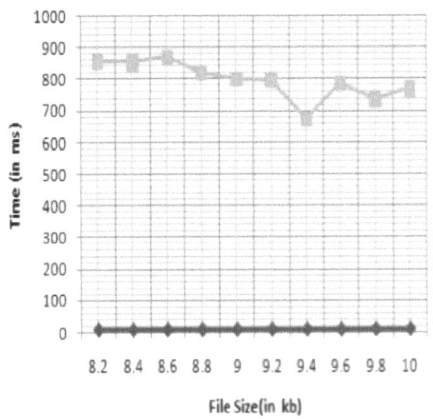

Fig. 3: Variation of Bandwidth Utilization (Source: self)

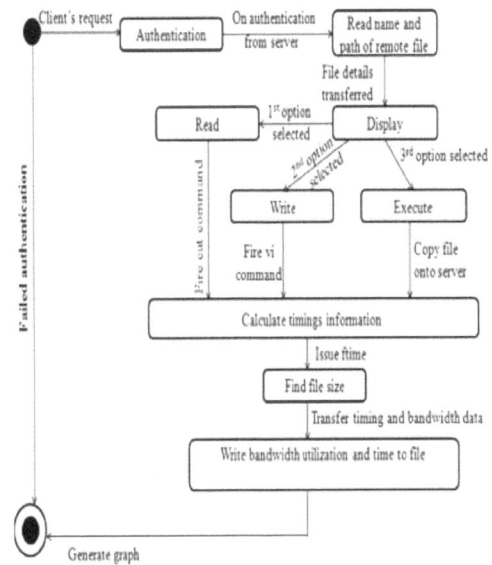

Fig. 2: State Transition Diagram

Linux server (Source: self)

ENHANCEMENTS IN NFS

Athicha Muthitacharen and her team[1] suggests an improved file update mechanism that prevents transferring iterative inessential data from different versions of a file over the constrained resources while providing substantial level of consistency. The suggested mechanism follows an open-to-close consistency model. It exploits inter-file similarities and breaks the files into "chunks". The chunks transferred are the ones that are not already present on the remote machine. The pivotal task, here, is to locate the exactly same chunks on both, the server and the client machines. For this, the SHA-1(Secure Hash Algorithm) hash function is used. The SHA-1 hash function provides different outputs to two input functions. This property of SHA-1 hash function is exploited by assuming the two chunks to be exactly same if they generate same output. To divide the file into chunks the technique of Rabin fingerprint is used which implements variable size shift-resistant blocks.

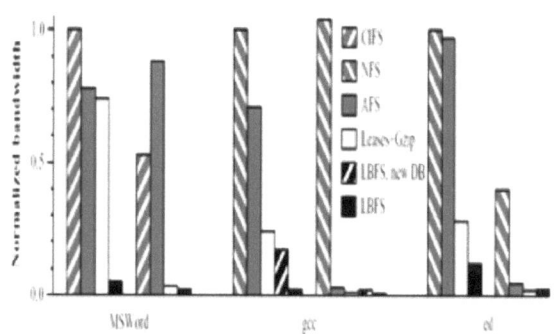

Fig. 4: Normalized bandwidth utilized by three workloads. The first four bars of each workload corresponds to upstream bandwidth, the second four bars to downstream bandwidth. The results are normalized against the upstream bandwidth of CIFS or NFS. (Source: Research Paper titled" A low bandwidth Network File System by Athicha Muthitacharoen)

PERFORMANCE OF IMPROVED NFS

In this section we concentrate on the bandwidth and network utilization of improved NFS under three common workloads and compare it to that of Common Internet File System(CIFS), NFS (native) and Andrew File System(AFS) as suggested in the paper titled 'A Low-bandwidth Network File System'.

We consider three cases namely:

1. MSWord File and make certain edits to it.
2. gcc, in this emacs 20.7 is recompiled from source.
3. ed, involves modifying the perl 5.6.0 source tree to change it into perl 5.6.1

The bar graph of Fig 4 shows that there have been substantial improvements.

PERCENTAGE SAVINGS IN THE THREE WORKLOADS:

WRITE Operation

MSWord (CIFS vs. Improved NFS)

(1.0-0.05) /1.0 *100 = 95%

MSWord (AFS vs. improved NFS)

(0.8-0.05) /0.8 *100 = 75%

READ Operation

MSWord (CIFS vs. Improved NFS)

(0.5-0.025) /0.5 *100 = 47.5%

MSWord (AFS vs. Improved NFS)

(0.9-0.025) /0.9 *100 = 87.5%

WRITE Operation

gcc (CIFS vs. Improved NFS)

(1.0-0.025) /1.0 *100= 97.5%

gcc (AFS vs. Improved NFS)

(0.7-0.025) /0.7*100= 96%

READ Operation

gcc (CIFS vs. Improved NFS)

(1.0-0.01)/1.0*100= 99%

gcc (AFS vs. Improved NFS)

(0.04-0.01)/0.04*100= 3%

WRITE Operation

ed (CIFS vs. Improved NFS)

(0.9-0.15)/0.9*100= 73%

ed (AFS vs. Improved NFS)

(0.89-0.15)/0.89*100= 83.14%

READ Operation

ed (CIFS vs. Improved NFS)

(0.35-0.05) /0.35*100= 85.74%

ed (AFS vs. Improved NFS)

(0.07-0.05)/0.05*100= 93%

IV. CONCLUSION

Network File System has been around for many years now. With increasing file and document sizes, network traffic has grown exponentially in the recent past. Hence, improving the performance of NFS is of prime importance.

Our results show the low bandwidth utilization of NFS but certain improvements show a drastic increase in performance. The improved version of NFS cuts down on the bandwidth utilized by exploiting the resemblance between files and techniques of compression and caching. For general operations such as editing documents and compiling software, the new improved version can consume significantly less bandwidth in comparison to the traditional file systems.

REFERENCES

[1] Muthitacharoen, B. Chen and D. Mazières, "A low-bandwidth network file system", *ACM SIGOPS Operating Systems Review*, vol. 35, no. 5, p. 174, 2001.

[2] Theodore Faber, "Optimizing Throughput in a Workstation-based Network File System over a High Bandwidth Local Area Network" SIGOPS Operating Systems Review, vol. 32, no. 1, pp. 29-40, January 1998

[3] Russel Sandberg, "The Sun Network Filesystem: Design, Implementation and Experience"

[4] Andrew S. Tanenbaum and Maarten Van Steen, *Distributed Systems: Principles and Paradigms* (New Jersey: Pearson Education Inc, 2007

A Review on Quad –Copter Fight Dynamics for Robotic Applications

Charu Khatri[1], Ayushi Chopra[1], Khyati Rai Jain[1], Pooja Gaur[1], Nisha Sinha[1], Akansha[1], Dr. Sneha kabra[1], Dr. Yogesh Pratap[1]

[1]Department of Instrumentation, Shaheed Rajguru College of Applied Sciences for Women, University of Delhi, Delhi, India

Abstract: In the last few decades Multi-copter has become a topic for research. Advanced researches have been made to make its applications in various areas like traffic surveillance, defence surveillance during strategic warfare, animal tracking in large natural habitats, aircraft inspections etc. This paper presents the design methodology and realization of the Quad-Copter, an aircraft based on a four-propeller design. This paper is about one among Arduino based Multi-copter containing four Motors i.e. A Quad-copter. A Quad-copter is aerodynamically unstable and requires on-board computer for stable flight. The main goal of this research work is to build an existing Quad-copter get a stable flight with landing and the camera to monitor.

Keywords:Quad-copter, propellers, rotor, roll, pitch, yaw, Arduino, UAV

I. INTRODUCTION

Quad-copters have a rich history. The first UAV was developed by Australian military to attack Venice, in Italy unmanned air balloons containing explosive but many of them reverted back to the Australian Army because they were unstable [1]. In 1920, Oemichen No 2 designed the first Quad-copter 4 rotors, 8 propellers and one motor, which was very stable for the time and had thousand successful flights [2]. In 1956 George De Brothezet and Ivan Jorme designed Quad-copter which uses thrust of four propellers in order to control roll, pitch and yaw. In recent years there has been a huge development, many small Quad-copters have entered the market including DJI Phantom and Parrot AR Drone. Quad-copter inhibits the qualities of different helicopter. Helicopters can be under pitched or co-axial while Quad-copters are comfortable blend of both. It is unmanned which follows pilot's instruction with the joystick on a remote control transmitter. Receiver receives and processes pilot's instruction which in turn moves the rotors and motors. Among the four rotors, two move clockwise and other two counterclockwise to negate any force or torque which finally stabilize the movement and function. Quad-copter has a centre

of gravity in middle of four rotors which keep its balance. Its existence for various purposes like military scientist in different countries use it for combat and reconnaissance, law enforcement Agencies use them in Search and rescue operations not only it can be used for aerial photography.

II. QUADCOPTER DYNAMICS AND CONTROL TECHNIQUE

Mathematicians came up with a way of describing

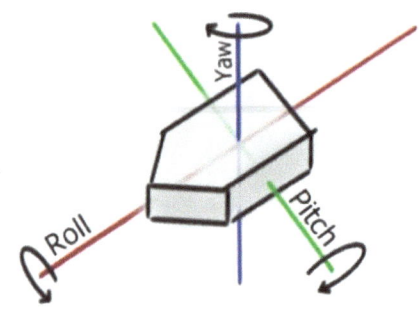

Fig 1: Dimensions of roll, pitch and yaw

the orientation of rigid bodies in space main use set of three angles to describe the orientation of multi rotor around the three spatial dimension they were called as roll, pitch and yaw.

The roll angle describes how the craft is tilted side-to-side. Rolling the Quad-copter causes it to move sideways. The pitch angle of the Quad-copter describes how the craft is tilted forwards or backwards. Pitching the Quad-copter causes it to move forwards or backwards. The yaw angle of the Quad-copter describes its bearing, or, in other words, rotation of the craft as it stays level to the ground. The root of the Quad-copter movements is the rotational speed of the Motors. By adjusting the relative speeds of the motors in just the right ways, keeping in mind that the rotational speed of the motors determines how much lift each prop produces, the flight controller is able to cause the Quad-copter to rotate around any of the directional axes (roll, pitch, and yaw), or make the Quad-copter gain or lose altitude. To make a Quad-copter roll right (or rotate about the roll axis clockwise), the flight controller will make the two motors on the left side of the Multi-rotor spin faster than the two motors on the right side. The left side of the craft will then have more lift than the right side, which causes the Multi-rotor to tilt.

Similarly, to make a Quad-copter pitch down (rotate about the pitch axis clockwise) the flight controller will make the two motors on the back of the craft spin faster than the two motors on the front. This

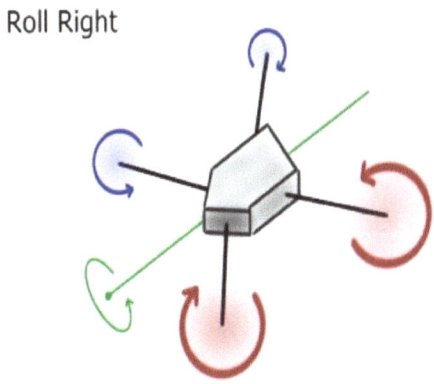

Roll Right

Fig 2: Roll movement

makes the craft tilt in the same way that your head

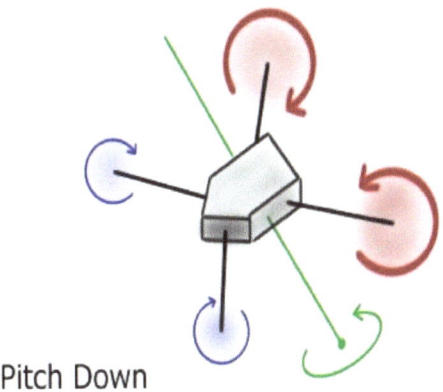

Pitch Down

Fig 3: Pitch movement

tilts when you look down.

Controlling the Quad-copter along the show yaw is quite difficult. We set up the Motors so that each Motor spins in opposite direction than its neighbors. For example if one motor is moving clock wise Quad-copter has a tendency to spin counterclockwise due to Newton's third law of motion which states that for every action there is an equal and opposite reaction.

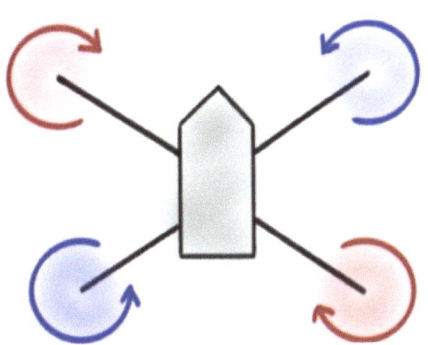

Fig 4: movement of motors

Designs of Quad-copter are divided into two stages that is part design in first stage and full interface at second stage. Flow chart of Quad-copter design is described in Figure below:

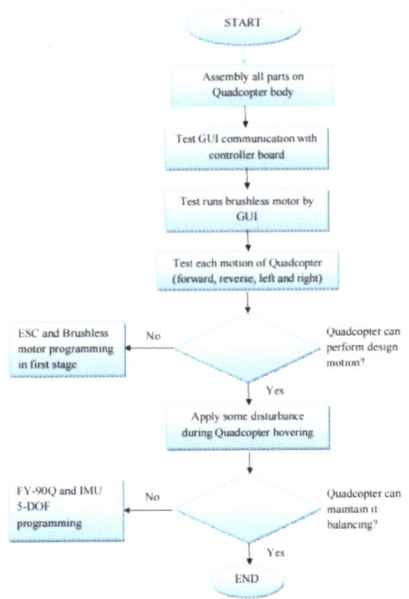

Fig 5: Flow Chart of QUAD-COPTER

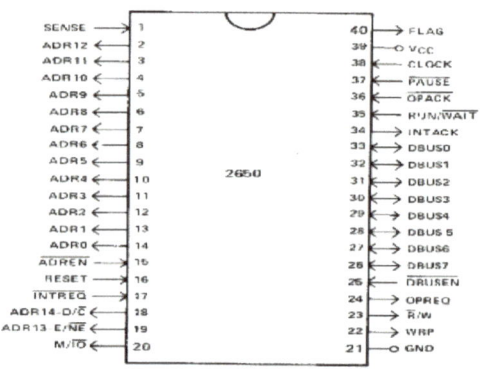

Fig 6: Pin configuration of Arduino 2560

COMPONENT USED

1.	Brushless Motor
2.	Propellers
3.	Frame
4.	Electronic Speed Controller
5.	NRF24L01
6.	Arduino MEGA 2560 Board
7.	Mounting Plate
8.	Flight Controller
9.	Power Distribution Board
10.	T plug
11.	Battery
12.	Voltage Regulator
13.	Camera

A2212 Brushless Motors:

No. of Cells:	2 - 3 Li-Poly 6 - 10 NiCd/NiMH
Kv:	1000 RPM/V
Max Efficiency:	80%
Max Efficiency Current:	4 - 10A (>75%)

No Load Current:	0.5A @10V
Resistance:	0.090 ohms
Max Current:	13A for 60S
Max Watts:	150W
Weight:	52.7 g / 1.86 oz
Size:	28 mm dia x 28 mm bell length
Shaft Diameter:	3.2 mm
Model Weight:	300 - 800g / 10.5 - 28.2 oz

PROCESSOR

The processor of the Arduino board uses the Harvard architecture where the program code and program data have separate memory. It consists of two memories such as program memory and data

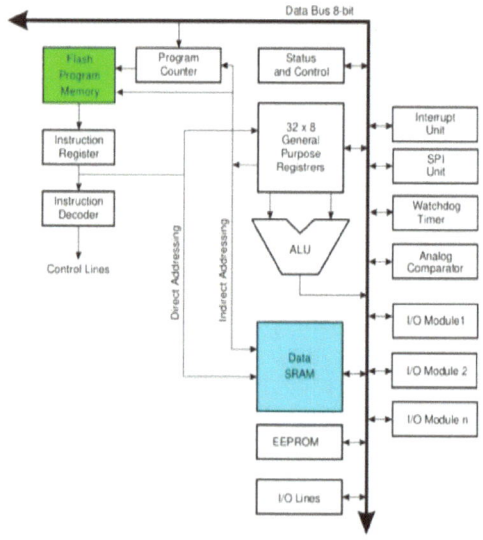

Fig 7: Block diagram of Arduino board

memory. Wherein the data is stored in data memory and the code is stored in the flash program memory.

III. METHODOOGY

QUAD-COPTER MOVEMENT MECHANISM

Quad-copter can described as a small vehicle with four propellers attached to rotor located at the cross frame. This aim for fixed pitch rotors are use to control the vehicle motion. The speeds of these four rotors are independent. By independent, pitch, roll and yaw attitude of the vehicle can be control easily.

Quad-copter have four inputs force and basically the thrust that produced by the propeller that connect to the rotor. The motion of Quad-copter can control through fix the thrust that produced. This thrust can control by the speed of each rotor.

TAKE-OFF AND LANDING MOTION MECHANISM

Take-off is movement of Quad-copter that landing position is versa of take-off position. Take-off (landing) motion is control by increasing (decreasing) speed of four rotors simultaneously, which means changing the vertical motion.

FORWARD AND BACKWARD MOTION

Forward (backward) motion is control by increasing (decreasing) speed of rear (front) rotor. Decreasing (increasing) rear (front) rotor speed simultaneously will affect the pitch angle of the Quad-copter.

LEFT AND RIGHT MOTION

For left and right motion, it can control by changing the yaw angle of Quad-copter. Yaw angle can control by increasing (decreasing) counter-clockwise rotors speed while decreasing (increasing) clockwise rotor speed.

HOVERING OR STATIC POSITION

The hovering or static position of Quad-copter is done by two pairs of rotors are rotating in clockwise and counter-clockwise respectively with same speed. By two rotors rotating in clockwise and counter-clockwise position, the total sum of reaction torque is zero and this allowed Quad-copter in hovering position.

IV. BENEFITS

Quad-copters can view all aspects of the planes without the need for special platforms

Quad-copters can get up close and personal with difficult to reach areas of the aircraft

Engineers can sit in centralized remote areas and review the photos and video.

V. CONCLUSION:

Quadcopters with extensive history and capabilities are the devices of the future. In this paper we are going to design quadcopter using quadcopter dynamics and control techniques. In outdoor test the quadcopter is going to drift from original position due to the influence of the airflows. This is going to control using a (GUI) Graphical User Interface.

REFERENCES

[1] the-history-of-drones-and-quadcopters, quadcopterarena.com

[2] https://en.wikipedia.org/wiki/Quadcopter

[3] Jared Rought, Daniel Goodhew, John Sullivan and Angel Rodriguez (2010). "Self-Stabilizing Quad-Rotor Helicopter".

[4] Allison Ryan and J. Karl Hedrick (2005). "A mode-switching path planner for UAV- assisted search and rescue." 44th IEEE Conference on Decision and Control, and the European Control Conference 2005

[5] A. Chamseddine, Y. Zhang, C. A. Rabbath, C. Join, and D. Theilliol, "Flatness-based trajectory planning/replanning for a quadrotor unmanned aerial vehicle," IEEE Transactions on Aerospace and Electronic Systems, vol. 48, no. 4, pp. 2832–2848, 2012.

[6] http://www.mwm.im/ResearchFiles/Papers/StabilityAndControlOfAQua drocopterDespiteTheCompleteLossOfOneTwoOrThreePropellers.pdf

[7] J. Fink, N. Michael, S. Kim, and V. Kumar, "Planning and control for cooperative manipulation and transportation with aerial robots," The International Journal of Robotics Research, vol. 30, no. 3, pp. 324–334, 2011.

Modeling and Simulation of Solar Photovoltaic Panels using LabVIEW Software

Ichettira, Shiva Kuttappa
Dept. of Electronics & Communication
Delhi Technological University
Delhi, India
Email: kshiv1996@gmail.com

Paramita Guha
Scientist
CSIR-CSIO, Delhi Centre
Delhi, India
Email: paramita.guha@csio.res.i

Abstract—In this paper, a solar panel is simulated using LabVIEW software. Four numbers of solar cells are connected in parallel Each solar cell is made of silicon and simulation is performed under different solar radiations. The output current and power are obtained for varying voltage. It has been observed the outputs improve significantly with the increase in number of solar cells. Hence, the given network can be considered as beneficial compared to the existing models available in literature. Numerical results along with discussions are presented. The simulation results are verified with another software.

Keywords—solar cell; solar panel; LabVIEW; photo-current; dark current component

I. INTRODUCTION

Solar power is gaining popularity throughout the world. Governments all around the world keep introducing new programs, which aim at popularizing solar power in their respective economies [1]. The popularity of solar power can be attributed to the fact that it is a clean and abundant source of energy, which can serve as a viable alternative to fossil fuels in the future. One of the most popular ways to harness solar power is using an array of solar cells. A solar cell or photovoltaic cell is an electrical device that converts the energy of light directly into electricity by employing photovoltaic effect.

The "photovoltaic effect" is the basic physical process through which a solar cell converts sunlight into electricity. This effect is the process in which two dissimilar materials in close contact produce an electrical voltage when struck by light or other radiant energy. Light striking crystals such as silicon or germanium diodes exhibit this property

the best and are economically viable which is why they are used as photovoltaic cells.

However, before actually designing a solar panel, it is required to have basic idea about its output waveforms due to application of various inputs. Hence, they are simulated before being actually implemented in hardware to ensure correct wiring and understanding of the circuit. Solar panels can be simulated in a large number of software, the most popular being the MATLAB software [2]-[4]. However, in this work, we have used LabVIEW to simulate our solar panel since it is user friendly, easy to understand and efficient. The main aim of this paper is to simulate a solar panel using LabVIEW software. The solar panel consists of four silicon solar cells. The solar cells are connected in parallel to each other to get maximum current output. The input for this network is taken as current whereas outputs are voltage and power. The paper is organized as the following. The Section II discusses the basic theory and problem formulation of the work. The simulation model is given in Section III. The results and discussions are given in the following section. Finally, the paper is concluded with Section V.

II. PROBLEM FORMULATION

The pictorial view of a simple solar cell is shown in Fig. 1.

When light is incident on the photo-diode, a photo-current I_{ph} is produced. This current is split into two, one is the current I_D drawn by the diode itself and the other is current I which flows into the external circuit. Hence, I_{ph} is expressed as a current source. The current I is the current available to us and is used to power other appliances, circuits and devices. Thus,

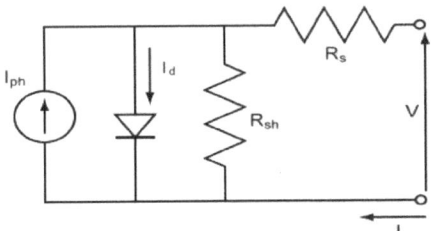

Fig 1: Connection diagram of a typical solar panel.

$$I = I_{ph} - I_D$$

(1)

In an ideal circuit, the shunt resistor R_{sh} and series resistance R_s in Fig. 1 would not exist. They are internal parasitic resistances which reduce power dissipation. Thus, R_{sh} would be infinite and R_s would be zero in an ideal circuit. Hence, the ideal circuit equation can be given as

$$I_D = I_{OC}(e^{\frac{qV}{kT}} - 1) - I_D$$

(2)

where I_{oc} is the saturation or dark current of the diode, q is the electron charge, V is the measured cell voltage, K is the Boltzmann constant and T is the temperature.

However, for a non-ideal diode, an ideality factor n, shunt resistance R_{sh} and series resistance R_s have to be considered. The ranges of R_{sh} is considered as 10 to 1000 of Mega ohms [5]. Shunt resistance is used to determine the noise current in the photo-diode with no bias (photovoltaic mode). For best photo-diode performance, the highest shunt resistance is desired.

The series resistance R_s of a photo-diode arises from the resistance of the contacts and the resistance of the undepleted silicon. It is given as

$$R_s = (W_s - W_d)\left(\frac{P}{A}\right) + R_c$$

(3)

where W_s is the thickness of the substrate, W_d is the width of the depleted region, A is the diffused area of the junction, P is the resistivity of the substrate and R_c is the contact resistance. The series resistance is used to determine the linearity of the photo-diode in photovoltaic mode with no bias, $V=0$. Although an ideal photo-diode should have no series resistance, typical values ranging from 10 to 1000 ohms are measured [5]. Hence, the equation of a non-ideal diode can be given as [5]

$$I = I_{ph} - I_{OC}\left(e^{\left(\frac{V+IR_s}{nKT}\right)} - 1\right) - \frac{(V + IR_s)}{R_{sh}}$$

(4)

with n being ideality factor.

III. SIMULATION

To simulate the solar panel on LabVIEW, we have used the following steps and also shown by the Flowchart in Fig. 2.

First, we will input the parameters by creating a control type constant for I_{ph} and I_{oc}.

Create numerical constant of value 1.6 x 10^{-19} for q. Create control type constants for n and T and multiply them. Now create a constant of value 91.38 x 10^{-23} for K. Multiply it with output of nT. We get nKT. Divide it by q.

Divide I_{ph} by I_{oc}. Increment it by 1. Now pass this expression through a log function box. We get $log_e((I_{ph}/I_{oc})+1)$.

Multiply the above two expressions to get V_{oc} [5].

Next, we insert a *for* loop which will start with $j=1$ and increment each time the loop is completed till it reaches 100. Inside the loop we will implement the $V=jV_{oc}/100$ to divide the V_{oc} into 100 divisions. Now, we implement the Eq. 4.

Next, create control type constants for R_s and R_{sh}. Multiply R_s with I. Add it to the V and multiply with q. We get $q(V+I.R_s)$. Divide it by nKT. Decrement it by 1, pass the resulting expression through an exponential function. Multiply this final expression with I_{oc}. We get $I_{OC} \cdot e^{\left(\frac{V+IR_s}{nKT}\right)} - 1$.

Now divide $(V+I.R_s)$ by R_{sh}. Subtract this from the above expression. We get I_D. Subtract I_D from I_{ph}. We finally get current I. Now multiply I and V. This product is the power P.

Hence, we create a control type constant which will denote the number of solar cells. We multiply this value with all the parameters: I, V and P.

We now feed the output of these three products into three arrays each of one-dimension. The values of each of the arrays of voltage, current and power are fed into three separate files for storage. The circuit is now complete.

Select the values and plot the \(P-V\) graph and \(I-V\) graph by inputting the values in MATLAB. Following these steps, the simulation circuit can be drawn and given in Fig. 3.

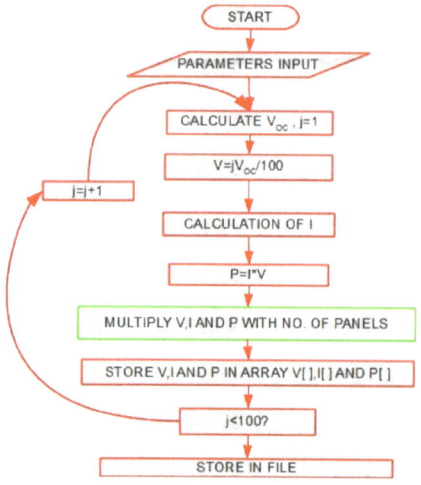

Fig 2: Flowchart of the Simulation Model.

IV. RESULTS AND DISCUSSION

Our circuit is now complete. We shall now input some values as an example to show simulation and check our circuit. After simulation, we shall follow steps 13, 14 and the outputs are obtained. The following parameters are given as inputs in the LabVIEW circuit [6], [7]: $T = 298\ K$, $n = 2.8$, $I_{OC} = 0.0008\ A$, $I_{ph} = 2.5\ A$, $R_s = 0.02$ ohm, $R_{sh} = 2000$ ohm, number of panels is taken as four. After simulation, the following results are obtained.

From Fig. 4, it is observed that as the voltage is increased, the current is maintained at 10 A and after a certain value of voltage, i.e. approximately 1.5V, the current starts decreasing and goes to zero. The relation between power and voltage is shown in Fig. 5. The power increases linearly with voltage and after approximately 1.8V, it starts decreasing and settles at $P= 0$.

In [5], almost same technique is used to simulate

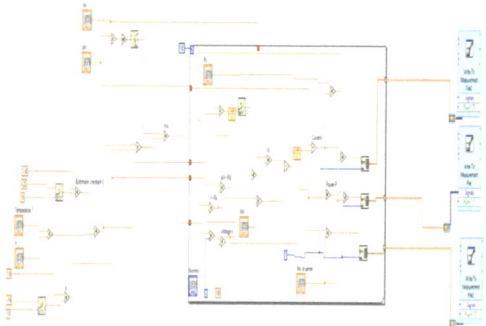

Fig 3: Block diagram of the LabVIEW model.

another solar cell. However, in that paper only one solar cell was used. Hence, the generated power and current was small. In our work, it has been shown that the current and power can be improved with more cells connected in parallel and better utilization of the renewable energy can be done.

It may be noted that we have taken practical values for a solar panel made up of four solar cells. The parameters, such as temperature and n can be changed depending on the environment. The resistances, R_s and R_{sh} can be replaced depending on availability. The current I_{ph} is dependent on solar intensity and I_{OC} will change depending on the components of the diode. Although the number of solar cells can be increased to improve the outputs, but some optimization technique can be used to limit the number of solar cells to obtain the desired output. This will be our future scope of study.

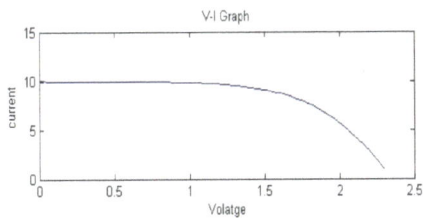

Fig 4: The output I-V characteristics

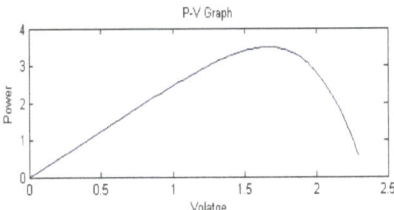

Fig 5: The output P-V characteristics

V. CONCLUSION

In this paper, we simulated a solar panel made of four solar cells at a temperature of 298 K and with an ideality factor of 2.8. The advantage of this circuit is that we can change the values of the parameters to simulate solar panels made up of different number of solar cells at different conditions up to a suitable limit. We also obtain a suitable range of values for current and power, which can later be optimized using MPPT (maximum power point tracking system) to increase efficiency. Hence, we obtain an efficient system, which can use solar energy.

REFERENCES

[1] S. P. Sindhu, V. Nehra, S. Luthra, "Recognition and prioritization of challenges in growth of solar energy using analytical hierarchy process: Indian outlook", *Energy*, vol. 100, pp. 332-348, 2016.

[2] J. Walker, Geoff, "Evaluating MPPT converter topologies using a matlab PV model", *Aust. J. Electr. Electron. Eng.*, vol. 21, no. 1, 2001.

[3] H. Atlas, M. Sharaf, "A fuzzy logic power tracking controller for a photovoltaic energy conversion scheme", *Electr. Power Syst. Res.*, vol 25, pp. 227-238, 1992.

[4] H. Belliaa, R. Youcefb, M. Fatimab, "A detailed modeling of photovoltaic module using MATLAB", *NRIAG Journal of Astronomy and Geophysics*, vol. 3, no. 1, pp. 53-61, 2014.

[5] G. Yang, M. Chen, "LabVIEW Based Simulation System for the Output Characteristics of PV Cells and the Influence of Internal Resistance on It", *ICIE '09 Proceedings of the 2009 WASE International Conference on Information Engineering*, vol. 1, pp. 391-394, 2009.

[6] http://www.pveducation.org/pvcdrom/solar-cell-operation/open-circuit-voltage.

[7] http://www.ni.com/white-paper/7230/en/.

High Performance Low Voltage Structures Of Operational Amplifier

Abhilasha Gupta, Jasdeep Dhanoa

agupta1810@gmail.com

Dept. of Electronics and Communication, IGDTUW, Delhi, India

Abstract: The basic two stage Operational Amplifier is revisited and two structures are proposed. One structure uses standard CMOS and DTMOS while the other is having DTMOS only. These two proposed structures are compared with the basic CMOS structure, operating at 1.8V power supply. The small signal analysis of all the circuits are computed and compared. They are also compared for the gain, PM, ICMR, CMRR, PSRR. All structures are verified through pspice simulations for 0.18um technology. The power dissipation of the proposed structures are 18μW and 0.6μW respectively, while that for standard CMOS is computed as 222μW.

I. INTRODUCTION

Operational amplifier is the most common building block of almost all electronic systems. This paper introduces two structures of op-amp that has considerable DC gain and low power dissipation.

The basic structure of CMOS based two stage Op Amp, is shown in figure 1. First stage is formed by transistors M1, M2, M3, M4 which provides the differential to single ended conversion. Second stage consists of current sink load inverter consisting of transistors M6 and M7 which provides additional gain in the amplifier [4]. The biasing of the Op Amp is achieved with the help of simple current mirror bias string and the transistors M5, M7 and M8 which make sure that the transistors are operating in the saturation region.

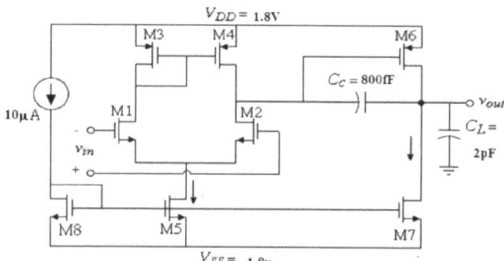

Figure 1: Block diagram of two stage Op Amp [4]

II. DYNAMIC THRESHOLD MOSFET (DTMOS)

Dynamic threshold MOSFET (DTMOS) transistor utilizes dynamic body bias; the gate and substrate of simple MOSFET are joined or tied together. Source substrate junction is forward biased by input gate voltage, thus ON stage threshold voltage (Vth) decreases and for OFF stage i.e. (gate is turned off), Vth returns to its original high value in equilibrium [2]. The symbol and transistor of DTMOS is shown in figure 2. The following equation shows the above fact.

$$V_{TH} = V_{TO} + \gamma \left(\sqrt{(f_o - V_{SB})} - \sqrt{f_O} \right) \quad (1)$$

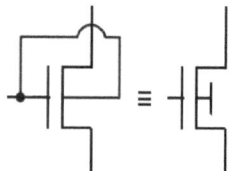

Figure 2: DTMOS SYMBOL

PROPOSED OPERATIONAL AMPLIFIERS

Two op-amp structures have been proposed working at ±0.9 V. The first structure is hybrid of CMOS and DTMOS (HYop-amp, Fig. 3) and the second structure is full DTMOS technology (DTop-amp, Fig. 4).

In figure 3 the first stage i.e. differential amplifier stage is designed with DTMOS, which reduces

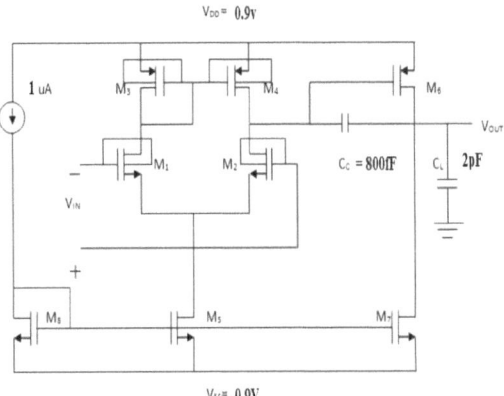

Figure 3: First Proposed Design: Hybrid amplifier: HYop-amp

threshold and power dissipation of the structure.

The second proposed structure (Fig. 4) has both the stages of op-amp replaced by DTMOS, which further reduces the power dissipation.

II. ANALYSIS OF AMPLIFIERS

Transfer function of basic CMOS op-amp (Fig.1) is given by equation (1) where, gm_1 and gm_2 are the input transconductance offered by the first stage and stage 2 respectively. R_1 and R_2 are the output resistances offered by stage 1 and 2 respectively. C_1 is the parasitic capacitance of stage 1 and C_L is the load capacitor. C_c is the compensation capacitor (miller capacitor) [4].

The transfer function is [4]

$$\frac{V_{out}}{V_{IN}}$$
$$= \frac{gm_1 gm_2 R_1 R_2 \left(1 - \frac{sC_C}{gm_2}\right)}{s^2 R_1 R_2 (C_1 C_L + C_2 C_c + C_c C_L) + s(R_2(C_C + C_L) + R_1} \quad (2)$$

The general transfer function is given by

$$\frac{V_{out}}{V_{IN}} = \frac{A_V \left(1 - \frac{s}{Z_1}\right)}{\left(1 - \frac{s}{p_1}\right)\left(1 - \frac{s}{p_2}\right)} \quad (3)$$

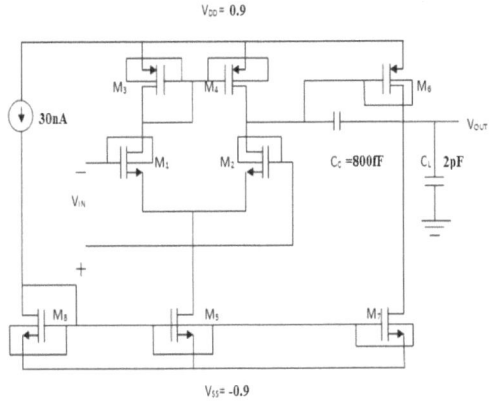

Figure 4: Second Proposed Design: Full DTMOS Amplifier: DTop-amp

Thus, zeros and poles of given system are given follow

$$Z_1 = \frac{gm_2}{C_C} \quad (4)$$

$$(4.20)$$

The dominant pole is located at

$$p_1 = \frac{-1}{gm_2 R_2 R_1 C_c}, p_2 = \frac{-gm_2}{C_L} \quad (5)$$

p_1 is the dominating pole

Gain of overall opamp is

$$A_V = gm_1 gm_2 R_1 R_2 \quad (6)$$

Gain bandwidth (GBW) is given by

$$GBW = A_V p_1 = \frac{gm_1}{C_C} \quad (7)$$

The small signal analysis of HYop-amp is given in figure 5. It is found that the first stage has an additional transconductance gm_b because of the body terminal acting in DTMOS, due to which there

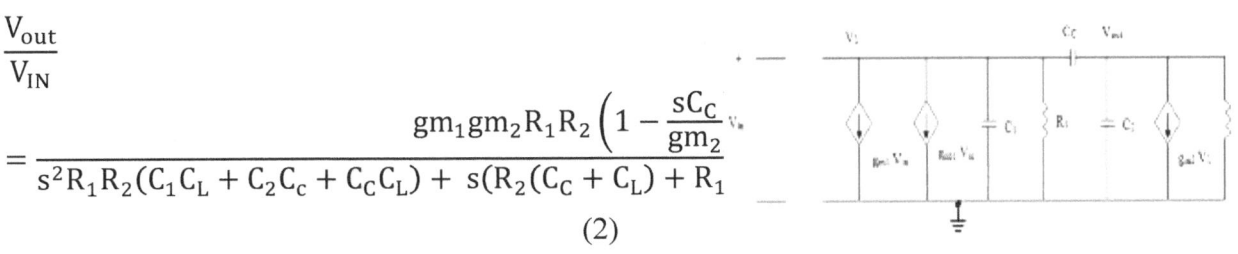

Figure 5: Small Signal Analysis of HYop-amp

is increase in gain

Node equation at V_1

$$\frac{V_1}{\frac{1}{sC_1}} + \frac{V_1}{R_1} + gm_1V_{IN} + gm_{b1}V_{IN} + \frac{V_1-V_{out}}{\frac{1}{sC_C}} = 0$$

(8)

Node equation at V_{out}

$$\frac{V_{out}-V_1}{\frac{1}{sC_C}} + \frac{V_{out}}{R_2} + \frac{V_{out}}{\frac{1}{sC_2}} + gm_2V_1 = 0 \qquad (9)$$

(4.39)

Using these 2 equations we get

$$\frac{V_{out}}{V_{IN}} =$$

$$\frac{(gm_1+gm_{b1})gm_2R_1R_2(1-\frac{sC_C}{gm_2})}{s^2R_1R_2(C_1C_L+C_2C_C+C_CC_L)+s(R_2(C_C+C_L)+R_1(C_C+C_1)+gm_2R_1R_2C_C)+1}$$

(10)

Now comparing this equation with general equation of transfer function (3)

$$A_V = (gm_1 + gm_{b1})gm_2R_1R_2 \qquad (11)$$

$$Z_1 = \frac{gm_2}{C_C} \qquad (12)$$

(4.42)

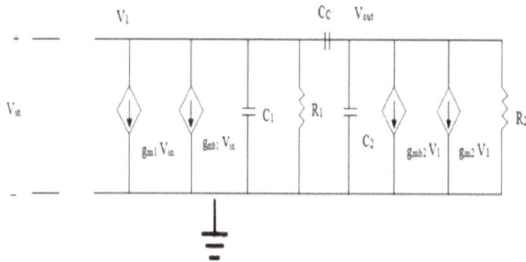

Figure 6: Small Signal Analysis of DTop-amp

The dominant pole (p_1) is located at

$$p_1 = \frac{-1}{gm_2R_2R_1C_C},$$

$$p_2 = \frac{-gm_2}{C_L} \qquad (13)$$

Thus, $GBW = A_Vp_1 = \frac{(gm_1+gm_{b1})}{C_C}$ (14)

On comparing the equation (6) and (11) it is analysed that the gain has increased, hence the gain bandwidth product.

The small signal analysis of DTop-amp is given in figure 6 in which the first stage has an additional transconductance gm_{b1} and gm_{b2} at second stage because of the body terminal acting in DTMOS, due to which there is increase in gain.

Node equation at V_1

$$\frac{V_1}{\frac{1}{sC_1}} + \frac{V_1}{R_1} + gm_1V_{IN} + gm_{b1}V_{IN} + \frac{V_1-V_{out}}{\frac{1}{sC_C}} = 0$$

(15)

Node equation at V_{out}

$$\frac{V_{out}-V_1}{\frac{1}{sC_C}} + \frac{V_{out}}{R_2} + \frac{V_{out}}{\frac{1}{sC_2}} + gm_2V_1 + gm_{b2}V_1 = 0 \qquad (16)$$

(4.47)

Using these 2 equations we get

$$\frac{V_{out}}{V_{IN}} =$$

$$\frac{(gm_1+gm_{b1})gm_2R_1R_2(1-\frac{sC_C}{gm_2})}{s^2R_1R_2(C_1C_L+C_2C_C+C_CC_L)+s(R_2(C_C+C_L)+R_1(C_C+C_1)+(gm_2+gm_{b2}).}$$

(17)

Now comparing this equation with general equation of transfer function of equation (3)

$$A_V = (gm_1 + gm_{b1})(gm_2 + gm_{b2})R_1R_2$$

(18)

$$Z_1 = \frac{(gm_2+gm_{b2})}{C_C} \qquad (19)$$

The dominant pole (p_1) is located at

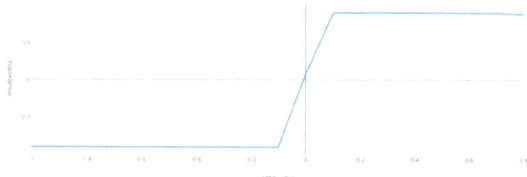

Figure 7: DC Characteristic of HYop-amp

$$p_1 = \frac{-1}{(gm_2+gm_{b2})R_2R_1C_C} \qquad (20)$$

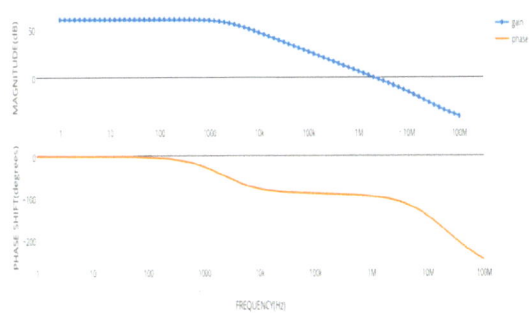

Figure 8: AC Characteristic of HYop-amp

$$p_2 = \frac{-(gm_2 + gm_{b2})}{c_L}$$

$$GBW = A_V p_1 = \frac{(gm_1 + gm_{b1})}{c_C} \qquad (21)$$

We can see from above analysis that proposed design have better gain and gain bandwidth product

III. RESULTS AND SIMULATION

The DC characteristics for HYop-amp is shown in

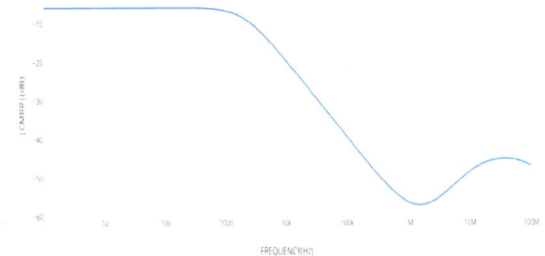

Figure 9: CMRR of HYop-amp

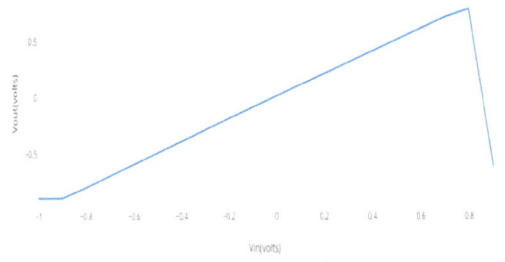

Figure 10: ICMR of HYop-amp

Fig.7 having offset value of 77mV. The bandwidth is calculated as 2 MHz for unity gain with

GAIN=60db and Phase margin = 102 deg shown by figure 8.

Figure 9 shows CMRR of HYop-amp with common gain -6.87 dB, with CMRR 66.87dB. Figure 10 gives ICMR with value of +-0.7V. Figure 11 and 12 gives PSRR$^+$ and PSRR$^-$ as 57.5 and 61 dB respectively

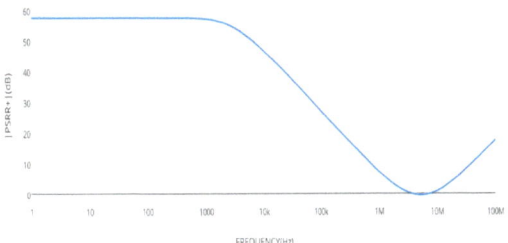

Figure 11: PSRR$^+$ of HYop-amp

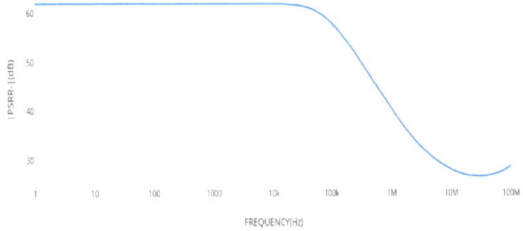

Figure 12: PSRR$^-$ of HYop-amp

DTop-amp is simulated with 30 nA bias current and +-0.9 volt supply. DC characteristics with 0.138 V offset value is shown in Fig.13. Figure 14 shows the AC characteristics, CMRR, ICMR and presented in fig 15 and 16 respectively. Figure 17 and 18 gives PSRR$^+$ and PSRR$^-$ as 66 and 78.6 dB respectively.

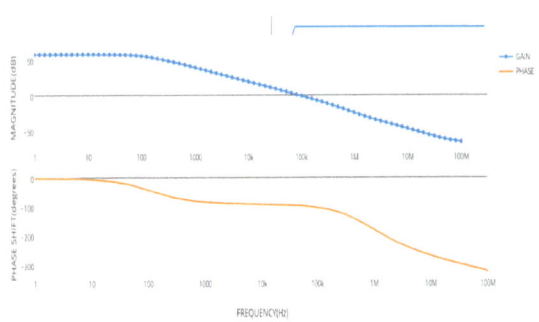

Figure 13: DC Characteristic of DTop-amp . Figure 14: AC Characteristic of DTop-amp

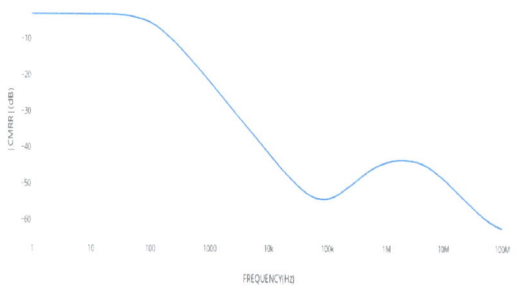

Figure 14: CMRR of DTop-amp

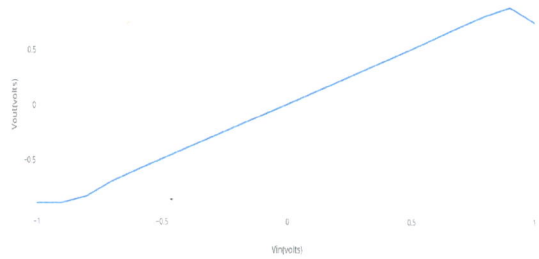

Figure 15: ICMR Characteristic of DTop-amp

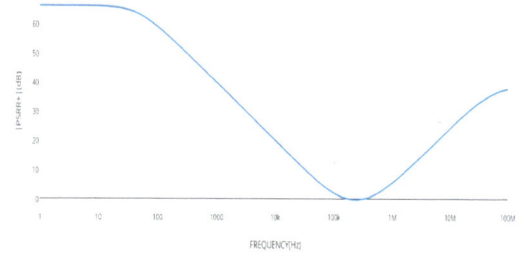

Figure 16: PSRR$^+$ Characteristic DTop-amp

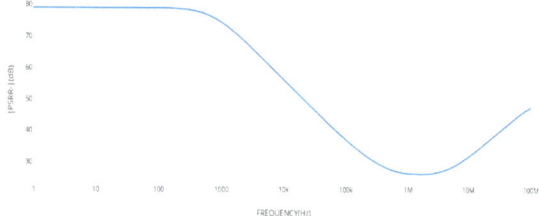

Figure 17: PSRR$^-$ Characteristic of DTop-amp

Table 1: COMPARISON OF THE THREE IMPLEMENTED DESIGN

Characteristics	CMOS op-amp (Fig. 1)	HYop-amp (Fig.3)	DTop-amp (Fig.4)
Voltages	+-1.8V	+-0.9V	+-0.9V
Bias Current(A)	10 u	1 u	30 n
DC(OFFSET)	0.04V	77Mv	.138v
Gain(dB)	60	60	58.2
Phase Margin(deg)	74	78	78
GBW	21 M	2 M	100K
CMRR(dB)	63.87	66.87	61.7
Power Dissipation uW	222	18	0.6
ICMR(V)	+1.6 - 0.8	-0.7, 0.7	0.8,- 0.7
PSRR+(dB)	65.23	57.5	66
PSRR-(dB)	81.69	61	78.6

IV. CONCLUSION

This paper presents and explores two op-amp structures. The Hybrid architecture (CMOS & DTMOS) reduces the threshold voltage and leakage currents, so it can be used for high performance applications. It also enhances CMRR and phase margin. Full DTMOS based op-amp structure is best for low power applications as it has a very low value of bias current and can be easily switched from saturation to linear region or vice versa by adjusting the bias current value.

REFRENCES

[1] Tanmeet K, Abhilasha G, Ruchi S, Jasdeep K, "Optimal Designs Of Application Specific Operational Amplifier", IEEE conference, 39th National System Conference, pp. 1-6, 2015

[2]   Assaderaghi F., Parke S. A., Sinitsky D., Bokor J., Ko P. K., Hu C., "Dynamic threshold-voltage MOSFET (DTMOS) for very low voltage operation", IEEE Electronic Device Letters, 15(12), pp. 510-512, 1994.

[3]   J.Mahattanakul, "Design Procedure for Two-Stage Cmos Opamp employing current buffer" IEEE trans. Circuits syst.II Fundam. Theory App, vol. 52, no.8, pp. 1508-1514, Nov 2005.

[4]   P. E. Allen, and D. R. Holberg, "CMOS analog circuit design", 2nd ed., New York: Oxford University Press, 2007.

[5]   Amana Yadav "International Journal of Engineering Research and Applications", (IJERA), Vol. 2, Issue 5, pp.647-654, September-October 2012.

[6]   R. Jacob Baker, "CMOS Circuit Design, Layout and Simulation", 2nd edition Wiley   Interscience, pp. 745-820, 2005.

[7]   Priyanka Kakoty, "Design of a high frequency low voltage CMOS Operational amplifier", International Journal of VLSI design & communication System (VLSICS), Vol.2, No.1, March 2011, pp. 73-85.

[8]   Jasdeep Kaur, S.S. Rajput and Nupur Prakash, "Self Cascode: A Promising Low Voltage Analog Design Technique", IETE Journal of Education, pp 73-83, vol 50, Issue 2, May-August 2009.

[9]   BehzadRazavi, Design of CMOS Analog Integrated Ckts, Mc-Graw Hill College, 2001. .

[10]

[11]   R. Sehgal, S.S. Rajput, S.S. Jamuar, "A 0.8V Operational Amplifier using Floating Gate MOS Technology " , Semiconductor Electronics, 2006 IEEE International Conference.

A review paper on Hollow Flashlight Thermoelectrically powered torch

Mehak Thareja, Kanika Juneja, Pinaakee Bhardwaj, Priyal Mangla, Ms. Himani Dua Sehgal, Ms. Reshma Sinha

Department of Instrumentation, Shaheed Rajguru College of Applied Sciences for Women

University of Delhi, New Delhi, India

Abstract — **Hollow flashlight runs solely on the heat of a human hand. The components of the flashlight are- LED lamp, peltier tiles, heat sink, IC LTC®3108, a capacitor, and a step up transformer. Thermoelectric effect is the basis for the construction of this flashlight. The thermoelectric effect deals with the direct conversion of temperature difference to electric voltage and vice versa. This voltage is generated by the temperature difference between the sides of the peltier tiles. Further voltage is converted to AC using IC LTC®3108 and is amplified by the step-up transformer to turn on the LED. A hollow aluminum tube serves as the heat sink.**

Keywords— *Thermoelectric effect, Heat Sink, Peltier Tiles, IC LTC®3108 , Stepup Transformer.*

I. INTRODUCTION

Humans are like walking 100 Watt light bulbs. From a sleeping bag that charges your gadgets to entire buildings warmed by body heat, scientists are harvesting the heat emitted by humans as a source of renewable energy. [1] With the aim of reducing the number of single-use batteries that are thrown in landfills, the motive is to develop an innovative flashlight, which can be designed cheaply and deployed to populations that can't afford electricity to light their homes.

The creation of thermoelectric flashlight will require the use of Peltier tiles (The Peltier tiles rely on temperature difference to transform heat from a human hand into energy to power an LED bulb that burns even brighter when it's cold outside.[1]), which produce electricity based on Peltier effect. In Peltier effect, temperature difference can be created between two junctions (p-n junctions) in order to get a current flow or voltage.[2] The current/voltage produced by the temperature difference will be DC in nature which needs to be converted into AC for further amplification. Hence, flashlight will also require a combination of IC LTC®3108 and step-up transformer.

With the LED across the transformer, a good LED brightness would be obtained with only 50mV DC input across oscillator.

Such technology will help prevent the unnecessary use of batteries, which leak toxic chemicals into the ground. And it is a cheap, renewable light source, which can be used by those that lack access to electricity. [1]

Design Of Body Heat Powered Light [3]

THE AVERAGE SURFACE AREA OF THE HUMAN SKIN IS 1.7 M^2 OR 17,000 CM^2, AS HUMAN DISSIPATES AROUND 350,000 JOULES PER HOUR, OR 97 WATTS SO THE HEAT DISSIPATION EQUALS TO 5.7MW/CM^2. A USEFUL AREA OF THE PALM IS ABOUT 10 CM^2. THIS IMPLIES THAT 57MW COULD BE AVAILABLE BUT ONLY 0.5 MW IS NEEDED TO GENERATE A BRIGHT LIGHT OF THE LED. THE DESIGN OF BODY HEAT POWERED LIGHT INCLUDES:

Peltier Tiles

Oscillator Circuit

Step-up Transformer

Heat Sink

Peltier Tiles

Peltier tiles are used to turn heat into electricity. They work on the principle of thermoelectric effect. The thermoelectric effect is the direct conversion of temperature differences to electric voltage and vice versa. [4] A thermoelectric device creates voltage when there is a different temperature on each side.[4] A Peltier cell also known as a thermoelectric cooler is made up of a large number of series-connected P-N junctions, sandwiched between two parallel ceramic plates.[3] The upper surface of the cell constitutes of a dielectric substrate. The peltier tiles are heavily doped. A peltier cell is also referred to as a thermoelectric generator. As a generator, it generates a dc output voltage, using seebeck effect, when the two plates are at different temperatures. The polarity of the temperature difference between the plates determines the polarity of the output voltage. The magnitude of the output voltage is proportional to the magnitude of the temperature difference between the Plates. [3] However, the output from a peltier is direct current. Direct Current cannot be multiplied, but if the DC is changed to AC, the voltage can be stepped up with a transformer. [3]

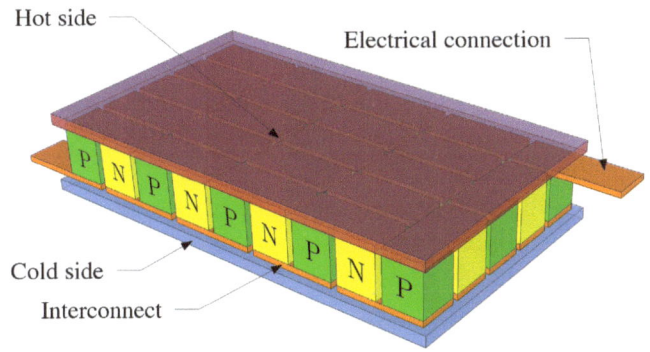

Figure 1[7] Peltier tile

Figure 2[6] Top view of LTC®3108

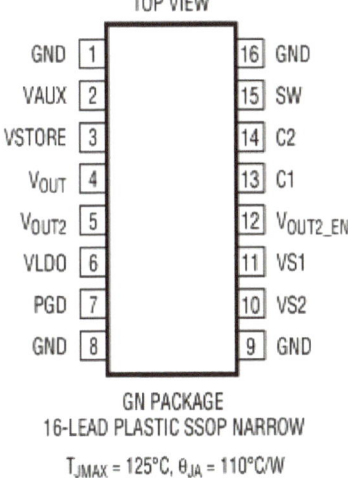

TOP VIEW

GND 1   16 GND
VAUX 2   15 SW
VSTORE 3   14 C2
V_{OUT} 4   13 C1
V_{OUT2} 5   12 V_{OUT2_EN}
VLDO 6   11 VS1
PGD 7   10 VS2
GND 8   9 GND

GN PACKAGE
16-LEAD PLASTIC SSOP NARROW

$T_{JMAX} = 125°C, \theta_{JA} = 110°C/W$

OSCILLATOR CIRCUIT

Direct current can be converted to alternating current using oscillators. The output from the Peltier is such a low voltage that needs to be boosted for which Linear IC LTC®3108 is used. [3] The

LTC®3108 is a highly integrated DC/AC converter ideal for harvesting and managing surplus energy from extremely low input voltage sources such as TEGs (thermoelectric generators), thermopiles & small solar cells [5]

The LTC®3108 can operate from a number of low input voltage sources, such as Peltier cells, photovoltaic cells or thermopile generators. The minimum input voltage required for a given application will depend on the transformer turns ratio, the load power required, and the internal DC resistance (ESR) of the voltage source.

The low voltage capability of the LTC®3108 design allows it to operate from a TEG with temperature differentials as low as 1°C, making it ideal for harvesting energy in applications in which a temperature difference exists between two surfaces or between a surface and the ambient temperature. The internal resistance (ESR) of most cells is in the range of 1Ω to 5Ω, allowing for reasonable power transfer. [6]

An external step-up transformer and a small coupling capacitor are used to form a resonant step-up oscillator which utilizes the MOSFET switch of LTC®3108. This allows it to boost input voltages as low as 20mV high enough to provide multiple regulated output voltages for powering other circuits. [3]

The frequency of oscillation is determined by the inductance of the transformer secondary winding and is typically in the range of 10 kHz to 100 kHz. For low input voltages, a primary-secondary turns ratio of about 1:100 is recommended. The AC voltage produced on the secondary winding of the transformer is boosted and rectified using an external charge pump capacitor (from the secondary winding to pin C1) and the rectifiers internal to the LTC®3108. The rectifier circuit feeds current into the VAUX pin, providing charge to the external VAUX capacitor and the other outputs. In application, a storage capacitor typically a few hundred microfarads is connected to V_{OUT} in. As soon as VAUX exceeds 2.5V, the V_{OUT} capacitor will be allowed to charge up to its regulated voltage.

The current available to charge the capacitor will depend on the input voltage and transformer turns ratio, but is limited to about 4.5mA. [3]

STEP-UP TRANSFORMER

The step-up transformer turns ratio will determine how low the input voltage can be for the converter to start. Using a 1:100 ratio can yield start-up voltages as low as 20mV. Other factors that affect performance are the DC resistance of the transformer windings and the inductance of the windings. Higher DC resistance will result in lower efficiency. The secondary winding inductance will determine the resonant frequency of the oscillator,

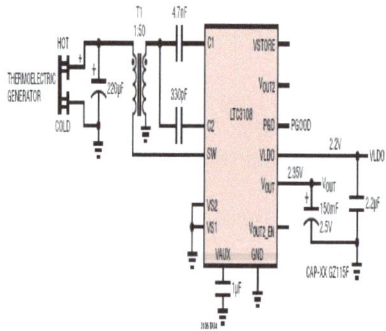

according to the following formula.

$$Frequency = \frac{1}{2 \cdot \pi \cdot \sqrt{L(sec) \cdot C}} Hz$$

Where L is the inductance of the transformer secondary winding and C is the load capacitance on the secondary winding. This is comprised of the input capacitance at pin C2, typically 30pF, in parallel with the transformer secondary winding's shunt capacitance. The recommended resonant (Peltier tile.) frequency is in the range of 10 kHz to 100 kHz.

HEAT SINK

Where the heat dissipation ability of the basic device is inadequate to bring down its degree of temperature, heat sinks play a vital role. Here, heat

sink allows the peltier tiles to cool down.

Usually aluminium is preferred to be used as heat sink, reason being its availability in abundance makes it cost effective. Also it provides high performance.

The peltier tiles are mounted on a hollow aluminium tube which is inserted in a larger PVC pipe, where the pipe has a cut that acts as an opening for the ambient air allowing it to cool the tube.

The hollow space facilitates easy air currents to flow freely through the tube. Due to this reason the flashlight is divided as two mediums: the outer area and the inner area of the tube.
Due to the contact between the petier and human hand, the outer area is also called as hot side. In contrary due to the passage of air currents inside the tube, the inner area is called as cold side.

II. METHODOLOGY

The following methodology has been taken in use:

1. Calculation/Estimation of amount of power required to obtain usable LED brightness (keeping in mind the heat dissipated by an average human being and also the thermal efficiency of Peltier tiles).
2. Testing the power generated by the Peltier Tiles on a per cm^2 basis.
3. Measuring area and internal resistance of the Peltier Tile.

4. Taping each tile onto a square aluminum tube.

One side of the Peltier is cooled by an ice pack, and the other side is heated with a 12-volt light bulb connected to a variable power supply.

The temperature difference between the sides of the Peltier tile is measured.

An LED is used to indicate the power generated (the voltage at which LED brightens is noted).

Approximately 2000-2500mV is needed to light a flashlight LED and according to a research, the voltage at which Peltier Tiles light an LED is much less. Since DC voltage cannot be amplified it is converted to AC using oscillator (IC LTC®3108). Afterwards this AC voltage is stepped up using a step-up transformer.

PHYSICAL FLASHLIGHT DESIGN

Peltier tiles are mounted on a milled area of aluminum tubing, and placed inside a larger PVC pipe, being insulated from it by air.

The tiles are hand griped through an opening in the PVC pipe.

The circuit is mounted in the front, and the LED centered in the middle of the tube.

The PVC pipe is wrapped with insulating foam.

III. CONCLUSION

Even with all the thermal and voltage conversion losses, there still remains enough power in the palm to provide enough voltage to light an LED. The results prove that some of the unused energy that could have been wasted in the form of heat is utilized to glow a flashlight using thermoelectric conversion by peltier tiles.

FUTURE ASPECTS

In the future, improving efficiencies of the converter, increasing the flashlight brightness, and perhaps using this technology for powering wireless medical sensors can be done.

The unique circuit and design has infinite possibilities and uses for the future. For example, heat from the metro seats having Peltier tiles, can be harvested and amplified into electricity using this method. Further, we can try storing this heat energy in order to power other devices.

In the field of Biomedical Instrumentation, for devices such as pacemaker implants that work on batteries that otherwise need to be replaced once in a while; with the concept of thermoelectric effect, we can try recharging the batteries of pacemaker using the body heat of the patient itself.

It is but, a means of discovering, what this concept, and what human heat energy, can do.

ACKNOWLEDGMENT
Authors are thankful to the Department of Instrumentation, Shaheed Rajguru College of Applied Sciences for Women, University of Delhi; for their support and co-operation.

REFERENCES

[1] http://www.inhabitat.com
[2] http://www.phy.olemiss.edu/~cremaldi/PHYS417/Seebeck%20and%20Peltier%20Effects.pdf
[3] B Ranjith Reddy et al Int. Journal of Engineering Research and Applications ISSN : 2248-9622, Vol. 4, Issue 8(Version 3), August 2014, pp.94-97
[4] Thermoelectric effect, Wikipedia
[5] http://www.linear.com/product/LTC®3108
[6] http://cds.linear.com/docs/en/datasheet/3108fc.pdf
[7] https://en.wikipedia.org/wiki/File:Peltierelement.png

Data-encryption techniques and algorithm for Watermarking on digital images using Fuzzy logic and its Significance

Rajshekhar Mukherjee

Assistant Professor, Maharaja Agrasen College, University of Delhi

Abstract: This paper illustrates the need for a security and encryption system to be present at the age of information boom, where millions of bytes of digital data are being exchanged over many digital platforms. Watermarking is a technique used commonly for safeguarding data without being perceptible to human vision, while placing the message within the content, thus protecting content even after decryption, file type conversion, compression or digital-to-analog conversion of data. Embedding the watermark in the form of noise, on a digital image, in frequency domain, while using fuzzy logic, can be a useful technique to encrypt the digital image file imperceptibly for copyright protection of owner, containing piracy across many channels and data hiding in modern communication systems. Detection of any change in the original image and reacquiring the watermark can be performed within region of acceptable fidelity and correlation coefficient.

Keywords- watermarking, embedding, hvs model of vision, fidelity, steganography,

I. INTRODUCTION

Watermarking is an important mechanism applied to physical objects like bills, documents, product packing etc. Physical objects can be watermarked using special dyes and inks or during paper manufacturing. Coded Images (Jpeg or Mpeg) and Compressed Audio signals can also be watermarked by means of binary watermarks or statistically distributed random numbers.

Watermarking does not necessarily hide the fact of secret transmission of information from third persons. Besides preservation of the carrier signal quality, watermarking generally has the additional requirement of robustness against manipulations intended to remove the embedded information from the marked carrier object. This makes watermarking appropriate for applications where the knowledge of a hidden message leads to a potential danger of manipulation. However, even knowledge of an existing hidden message should not be sufficient for the removal of the message without knowledge of additional parameters such as secret keys. A crucial feature of digital watermarking is to hide the additional information directly in the signal. This requires a certain perceptual threshold allowing the insertion of additional information and hence distortions of the carrier signal without incurring unacceptable perceptual degradation of the original carrier signal.

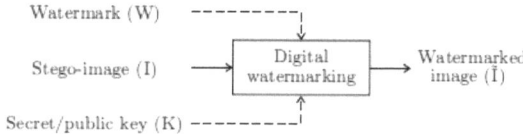

Figure: General digital watermark extraction/recovery scheme

The past few years have witnessed a rapid growth in the number and variety of applications of fuzzy logic (FL). FL techniques have been used in image-understanding applications such as detection of edges, feature extraction, classification, and clustering. Fuzzy logic poses the ability to mimic the human mind to effectively employ modes of reasoning that are approximate rather than exact. In traditional hard computing, decisions or actions are based on precision, certainty, and vigour. Precision and certainty carry a cost. In soft computing, tolerance and impression are explored in decision making. The exploration of the tolerance for imprecision and uncertainty underlies the remarkable human ability to understand distorted speech, decipher sloppy handwriting, comprehend nuances of natural language, summarize text, and recognize and classify images. With FL, one can specify mapping rules in terms of words rather than numbers. Computing with the words explores imprecision and tolerance. Another basic concept in FL is the fuzzy if–then rule. Although rule-based systems have a long history of use in artificial intelligence, what is missing in such systems is machinery for dealing with fuzzy consequents or fuzzy antecedents. In most applications, an FL solution is a translation of a human solution. Thirdly, FL can model nonlinear functions of arbitrary complexity to a desired degree of accuracy. FL is a convenient way to map an input space to an output space. FL is one of the tools used to model a multi-input, multi-output system.

Soft computing includes fuzzy logic, neural networks, probabilistic reasoning, and genetic algorithms. Today, techniques or a combination of techniques from all these areas are used to design an intelligence system. Neural networks provide algorithms for learning, classification, and optimization, whereas fuzzy logic deals with issues such as forming impressions and reasoning on a semantic or linguistic level. Probabilistic reasoning deals with uncertainty. Although there are substantial areas of overlap between neural networks, FL, and probabilistic reasoning, in general they are complementary rather than competitive. Recently, many intelligent systems called neuro fuzzy systems have been used. There are many ways to combine neural networks and FL techniques. In this paper, the FL concepts such as fuzzy sets and their properties, FL operators, hedges, fuzzy proposition and rule-based systems, fuzzy maps and inference engine, defuzzification methods, and the design of an FL decision system have been used.

The field of Digital image Processing refers to processing of digital images using digital computer. Digital image composed of finite number of elements called pixels, each of which has a particular location and value.

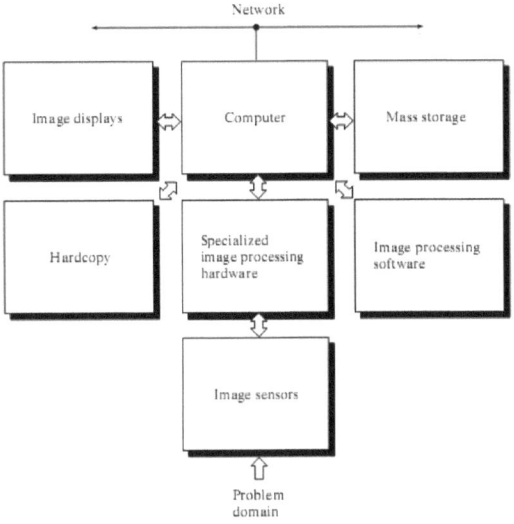

II. PROCEDURE

To embed a watermark on a digital image, the image is treated as a signal while the watermark acts as noise. For transforming the digital image from spatial-domain into frequency domain, one can use DCT (Discrete Cosine Transform). Spread-spectrum approach to watermarking is also used as it can be advantageous for transferring narrowband data through a noisy channel, by modulating each data symbol with a wideband and low amplitude signal.

HVS parameters luminance sensitivity, contrast and frequency sensitivity are significant to embedding an imperceptible watermark. Fuzzy Inference system is also used to weigh the HVS factors.

After embedding the watermark, inverse DCT is performed on the image and an algorithm to extract and detect the watermark is executed.

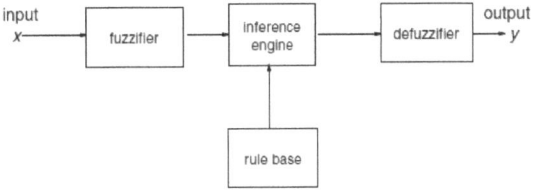

Fig: Fuzzy inference engine block process

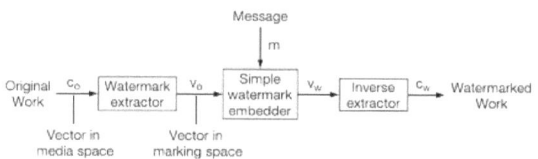

Fig: General three-step watermark embedding process

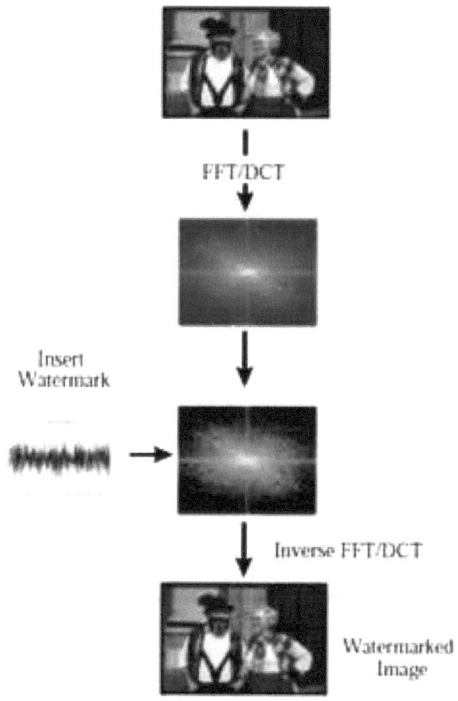

Fig: Watermark embedding in spectral domain

III. EXPERIMENTAL RESULTS, RELIABILITY ANALYSIS AND DISCUSSION

Fig: Original Image and Watermarked Image

Perceptible quality of the image before and after watermark embedding remains almost same in all the HVS –Fuzzy weighted watermarking scheme

which have been adopted in this paper. The three essential parameters used in this present work, namely Luminance sensitivity, Edge sensitivity and Contrast sensitivity, for each block is obtained and then are used to find the weighing factor, using a Fuzzy Inference System (FIS).This is done to embed the watermark onto host image which will be more robust as factors of the visual dynamics of our eyes and its limitations have been taken into account and utilized it to embed a watermark having minimum perceptibility and robustness simultaneously. The watermark generated is a vector of normalized random numbers with mean of zero and variance of one, N (0, 1) and embedded into the low DCT coefficients of each block using the formula mentioned in previous section. This can be called a hybrid form of watermark embedding.

Robustness of the watermarking procedure has been checked by making the signed image undergo various *StirMark* attacks during the course of the present work. To compute and ascertain the perpetual quality of the image watermarked with the host image, three quality checks were performed on images- MSE, PSNR, and Structural Similarity (SSIM).

The watermark extraction is based on the spread spectrum algorithm COX et al have given. The Similarity Correlation function, SIM(X,X*), is also computed for the original watermark after extraction as well as other randomly generated watermarks and seen to be highest by a large margin(better threshold) in case of the original extracted watermark only. Thus one can also verify, by the higher value of SIM(X, X*), that the watermark recovery process yields a good recovery of the watermark in the present work. Hence the algorithm followed in the present work can find practical uses in various communication application and transmission techniques in the field of data security.

IV. CONCLUSION

The watermark embedding with an algorithm which encompasses the Human Visual System (HVS) and hence makes the watermark more robust to the common image processing attacks as the human eye cannot see the difference in the perpetual quality of the original and signed images, has been done in our work and successfully tested, as indicated by the MSE, PSNR and SSIM values of the signed images obtained after performing the image extraction. These values, for different cases are within the expected range for these attacks. Also watermark embedding has been performed on coloured images using various colour channels and get the visual quality metric's values within the expected range.

The various attacks are performed on the signed image as commonly encountered in communication - jpeg compression, additional noise, Gaussian blur, Rotation by 90^0, brightness and contrast operation, Gaussian noise, and median filtering .

Also the SIM(X, X*) value obtained for the signed image watermark indicates a good watermark-recovery process, along with robustness of the watermark. Thus a hard decision detector based on

a threshold value is also implemented in the present work.

Using the technique of steganography algorithm along with watermarking in frequency domain with spread-spectrum approach will make the watermark difficult to detect and more robust to attacks.

REFERENCES

[1] Yeung, M. M.; Mintzer, F.; "An Invisible Watermarking Technique for Image Verification"; *Proc of ICIP 1997*, Santa Barbara, CA; pp. 680-682

[2] M. Kutter and F.A.P. Petitcolas, "A fair benchmark for image watermarking systems," in Proc. Electronic Imaging '99, Security and Watermarking of Multimedia Contents, vol. 3657. San Jose, CA, (Jan. 1999), pp. 226–239

[3] Cox, J.; Miller, M. L.; Bloom, J. A.; Fridrich J. & Kalker T. (2008). *Digital Watermarking and Steganography*, Morgan Kaufmann Pub., Elsevier Inc.

[4] Cox, J.; Kilian, J.; Leighton F. T. & Shamoon T. (1997). Secure spread spectrum watermarking for multimedia. *IEEE Transactions on Image Processing*, Vol. 6, No. 12,(December 1997),pp.1673-1687

[5]

[6] CRAVER, S., MEMON, N., YEO, B. L., AND YEUNG, M. M. 1998. Resolving Rightful Ownerships with Invisible Watermarking Techniques: Limitations, Attacks and Implications. *IEEE Journal on Selected Areas in Communications 16*, 4 (May), 573–586.

Innovative Teaching Strategy In Digital Age

Divya Gupta

Assistant Professor

Economics Department

Daulat Ram College

University of Delhi

Email- gupta.divya99@gmail.com

Abstract- Digital revolution has transformed the way young people communicate, network, seek help, access information and learn. Social networking websites, such as Facebook, Twitter and Youtube are an integral part of communication for today's college students. The purpose of this study is to examine how incorporating social media into teaching process can contribute to students' learning and engagement. Online survey was administered to a sample of college students (in the experimental group and the control group). A total of 80 first year Economics students of Daulat Ram College participated in this study (40 in the experimental group and 40 in the control group). With the experimental group, social media was used for various types of academic discussions and co-curricular activities. Engagement was measured using a 24-item scale based on the National Survey of Student Engagement. The two group's difference in engagement was analyzed using fixed effects analysis of variance (ANOVA) model (using SPSS software). The ANOVA results showed that the experimental group observed a significantly greater increase in engagement as compared to the control group. This study provides experimental evidence that Social Networking websites can be used as powerful tool for collaborative learning thereby improving teaching-learning process.

Keywords- collaborative learning, learning communities, media in education, digital technology

I. INTRODUCTION

New developments in the technological world have made the internet an innovative way for individuals and families to communicate. In fact, university students share a large percentage of the total number of users who engage in online social networking activities.

Students nowadays are tech savvy. There are 125 million facebook users in India with 59 million accessing it daily.

Incorporating social media in higher education can be beneficial to both students and instructors. Instructors will be able to improve their teaching methods and can stimulate students' thinking by posting messages or launching dialogs with specific students even after the normal class time. They can design and develop course related online social networking websites to enhance student learning. With such technology, students can discuss the study tips among themselves and with the instructors. Moreover, they can communicate and extensively share knowledge among the peers. Through this kind of intensive communications and knowledge sharing between instructors and students, and among the peer group, students' knowledge acquisition processes and their academic performance will be improved

II. SCOPE OF STUDY

McCarthy (2010) reported that in a 2009 study using *Facebook* as part of a blended learning approach, the vast majority (73-88%) of students felt that the Facebook group activities helped them develop academic and social relationships with their classmates [1]

There has been interest in integrating various social media tools (such as blogs, microblogs, video-sharing sites, and social networking) into the learning process [2] especially by faculty members with a disposition towards the use of newer technology in education [3].

Terry Anderson's (2008) model states that online education must be learner-, knowledge-, assessment-, and community-centered, and must make use of existing learning theories, but transform them to fit online contexts[4]. It is the transformation of content from traditional delivery modes to online social media that has stymied educators and researchers, who find that the social media sphere is difficult to integrate with educational objectives.

Rosenfield (2009) argued that resistance to using social media for education "prevents the effective use of technology in teaching, the acquisition of vital 21st-century literacy skills by students, and the professional development of teaching staff" [5]

Today, engagement is conceptualized as the time and effort students invest in educational activities that are empirically linked to desired college outcomes [6]. Engagement encompasses various factors, including investment in the academic experience of college, interactions with faculty, involvement in co-curricular activities, and interaction with peers [7]. Kuh (2009) emphasizes two major facets: in-class (or academic) engagement and out-of-class engagement in educationally relevant (or co-curricular) activities, both of which are important to student success.

Two recent studies have focused specifically on social media and engagement and have found relationships between time spent on social media and student engagement. Both the Heiberger and Harper (2008) and HERI (2007) studies found a positive correlation between social networking website use and college student engagement [8, 9].

Currently, the research on the use and benefits of incorporating online social networking sites into the educational process is limited in India to a few small studies .This project is first-of-its-kind that has been undertaken in India.

This study will address the following Research Hypothesis-

H1: Encouraging the use of social media for educationally relevant purposes has positive impact on students' engagement?

III. RESEARCH METHODOLOGY

DATA COLLECTION-

To address the research questions, data was collected from a non-random sample of students at Daulat Ram College, University of Delhi during the summer semester through an online survey in thecollege. The students were randomly assigned to

the experimental group and to the control group. The experimental group used social media for academic related activities while the control group did not.

Students were asked to participate in the study by taking a pre- and post-test survey. The questionnaire was uploaded and the link was sent to the participating students. Students took a survey prior to the start of the course to assess several factors including:

- Current online social networking habits, based on number of hours per week spent on websites such as Facebook, Twitter etc.
- Access to Internet-capable computers on campus and at home/in dormitories
- Access to social-networking websites via mobile devices

The following activities were done with the experimental group through social media in order to address the research questions-

- Organized study groups
- Connected students with each other and with instructors
- Provided assignments to students
- Academic discussions
- Provided information on academic and co-curricular activities on campus
- Class reminders
- Posting of links, notes, or videos related to course content.

At the end of the semester, students completed a survey to determine whether there had been changes in attitudes, values, and beliefs about social networking and its relationship to education, as well as the nature of these changes.

MEASURE

Engagement scale: Pre- test and post- test engagement scale. Engagement was measured using a 24-item scale based on the National Survey of Student Engagement.

The *Statistical Package for Social Sciences* (SPSS) was used to analyze the data for this study.

TESTING RESEARCH HYPOTHESIS

SOCIAL MEDIA AND ENGAGEMENT

In order to assess changes between the pre- and post- test measurement of engagement, we defined Difference in Engagement score=Post- test engagement score -pre-test score. Dependent variable is Difference in Engagement Score.

The differences in engagement between the two groups- control and experimental group was analyzed using fixed effects analysis of variance (ANOVA) model.

IV. ANALYSIS AND RESULTS

Table 1 shows the descriptive data for engagement score difference by groups. The control group (group 1) has mean engagement score 3.256 and standard error 0.85 whereas the experimental group has mean engagement score of 7.229 and standard error of 0.897. Table 2 shows that the experimental group had significantly (P value- 0.002) higher difference scores as compared to control group. We also conducted a fixed effects ANOVA model with pre-test engagement scores as the dependent variable and found that there were no pre-existing differences in engagement.

Estimates (Table 1)

Dependent Variable: Engagement

Group	Mean	Std. Error	95% Confidence Interval	
			Lower Bound	Upper Bound
1	3.256	.850	1.562	4.950
2	7.229	.897	5.440	9.017

Comparisons(Table 2

Dependent Variable: Engagement

(I) Group	(J) Group	Mean Difference (I-J)	Std. Error	Sig.
1	2	-3.972[*]	1.236	.002
2	1	3.972[*]	1.236	.002

V. ENGAGEMENT INDICATORS

Engagement Indicators
Collaborative Learning
Discussion with Diverse
Student-   Faculty Interaction
Reflective & Integrative Learning
Effective   teaching practice
Higher Order Learning
Quantitative Reasoning

VI. IMPLICATIONS

Our results suggest that Social Networking websites can be used to engage students in ways that enhances learning and encourages intellectual and overall development. Students collaborated through Facebook on projects assigned to them and developed interpersonal skills. They also shared their strengths and actively interacted with each other by sharing their experiences. Facebook also helped students to develop critical thinking, creating connections among the concepts and develop problem solving ability. It improved the interactions between the students and faculty members beyond the classroom and offered the deepest kind of learning experiences by enabling students to ask questions related to their own struggles and interests, to take responsibility for their own intellectual development, and to make more personal connections with their teachers. Moreover, it helped in forming study groups which allowed students to learn a lot more, in a lot shorter time in enjoyable manner. Lastly, Facebook helped in increasing interaction among the diverse groups, encouraged students to grow outside their boundaries and motivated the students to participate online who otherwise were not active in classroom.

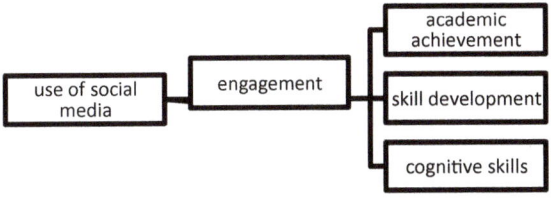

VII. CONCLUSION

The aim of this study was to examine whether it is effective to utilize online networking sites to design collaborative learning activities, e.g., exchange of project information, to increase interactivity and learning among individual students, and develop a rich environment for socialization. This study provided the experimental evidence that using social networking websites in educationally relevant ways can increase student engagement and that social media can be used as an educational tool to help students collaborate and facilitate their learning. The world of education is currently [10]).

undergoing a massive transformation as a result of the digital revolution and social media can play an important role. New technologies create learning opportunities that is taking their education out of school into homes, libraries, Internet cafes, where they can decide what they want to learn, when they want to learn, and how they want to learn. As there is continuing growth in the use of social media by college students and faculty, it is hoped that this study will motivate further studies of Facebook and other social media to optimally use emerging technologies for educational purposes.

REFERENCES:

[1] McCarthy, J. (2010), Blended learning environments: Using social networking sites to enhance the first year experience. Australasian Journal of Educational Technology, 26(6), 729-740.

[2] Grosseck G. & Holotescu C. (2009) *Can we use Twitter for educational activities?* Proceedings of the 4th International Scientific Conference: eLearning and Software for Education, Bucharest, Romania. Available at:http://adlunap.ro/eLSE_publications/papers/2008/015.-697.1.Grosseck%20 Gabriela-Can%20we%20use.pdf (last accessed 12 January 2010).

[3] Crook C. (2008) *Web 2.0 technologies for learning: The current landscape opportunities, challenges, and tensions.* Becta Research Reports. Becta, Coventry. Available at:http://research.becta.org.uk/uploaddir/downloads/page_documents/research/web2_technologies_learning. pdf (last accessed 7 January 2010).

[4] Anderson, T. (2008). "Towards and Theory of Online Learning." In Anderson, T. & Elloumi, F. *Theory and Practice of Online Learning.*

[5] Rosenfeld, E. (2008, February). Blocking Web 2.0 tools in schools: creating a new digital divide. *Teacher Librarian.* p. 6. Retrieved from EBSCOhost.

[6] Kuh G.D. (2009) What student affairs professionals need to know about student engagement. *Journal of College Student Development* 50, 683–706.

[7] Pascarella E.T. & Terenzini P.T. (2005) *How College Affects Students: A Third Decade of Research.* Jossey-Bass, San Francisco, CA.

[8] Heiberger G.&Harper R. (2008) Have you Facebooked Astin lately? Using technology to increase student involvement. In *Using Emerging Technologies to Enhance StudentEngagement. New Directions for Student Services Issue #124* (eds R. Junco&D.M. Timm), pp. 19–35. Jossey-Bass, San Francisco, CA.

[9] Higher Education Research Institute (2007) *College freshmen and online social networking sites.* Available at: http://www.gseis.ucla.edu/heri/PDFs/pubs/briefs/brief-091107-SocialNetworking.pdf (last accessed 7 March 2010

Architecture for an Internet of Things Collision Detection System

Nehal Sharma,
B.Tech Computer Science (4th year)
Shaheed Rajguru College of Applied Sciences for Women,
University of Delhi,
Email id: nehal.srcasw.du@gmail.com.

Megha Sahni,
B.Tech Computer Science (4th year)
Shaheed Rajguru College of Applied Sciences for Women,
University of Delhi,
Email id: msahni025@gmail.com

Abstract: Internet of Things (IoT) is a relatively new concept that has gained prominence in the past few years. While a number of applications have been proposed that make use of Internet of Things, we, in this paper aim to present a model for collision detection system using IoT. The conventional collision detection systems have been around for many years now but their capabilities are limited to warning only an individual vehicle or driver about an upcoming collision. The focus of this paper is directing the potential of a collision detection system to communicating with multiple objects at the same time through the use of IoT. We also shed light on the architectural components of such a system while highlighting the challenges that exist in this field.

Keywords: Internet of Things, V2V communication, Collision Avoidance System, Intelligent Vehicles.

I. INTRODUCTION

The revolutionizing term 'Internet of Things' (IoT) was first coined by Kevin Ashton. The year, 1999 was a massive year for IoT, as highly productive and efficient research was carried out in this field in the laboratories of Massachusetts Institute of Technology. It was the time when our lives got changed from the ways we drive to the ways we could think an electronic gadget could serve us. Internet of Things (IoT) is a means of integrating the physical objects and ensuring communication amongst them. Cisco [5] defines "Internet of Everything" as the "latest wave of the Internet—connecting physical objects… to provide better safety, comfort and efficiency" while IBM[5] describes it as "a completely new world-wide Web, one comprised of the messages that digitally empowered devices would send to one another. It is the same Internet, but not the same Web."

IoTs' applications such as smart cities, smart transportation, telemedicine, RFID's, logistics etc. can provide efficient and instant solutions to human problems. Use of IoT in automobiles has changed the definition of driving and comfortable commuting with the advancement of connected cars. At very low cost we can integrate the intelligent devices and can reap benefit from this connectivity by using data analytics that will extract the meaningful information and convert the automobile into a mini moving data centre.

Not only this, the technology has even made it possible to detect the accidents that may have occurred due to drivers' negligence or due to collision of cars due to poor weather conditions, through sensors, GPS and GSM technologies. To avoid the road accidents Collision Detection Systems are used. In case the accident has taken place, the information, including the accident location, can be sensed and sent to the relatives of the driver.

In this paper, we aim to propose a collision detection system which makes use of IoT. In such a system, the vehicles will be able to communicate with each other and also warn each other about a collision that is likely to happen. Smart devices like sensors, cameras and actuators can monitor the relative speed of vehicles which can then help to control the collisions.

II. LITERARY BACKGROUND

Previous works related to collision detection system in automobiles makes intensive use of sensors that employ radar, infrared or ultrasonic waves for detecting and calculating the proximity with the leading vehicles. RADAR sensors have been found to be more reliable because of their long detectable range [4]. Ultrasonic waves have also been used to detect objects that fall in the blind spots of the vehicle.

A Fuzzy based collision detection and avoidance system has been proposed by Jeich [2]. The model discussed in this paper talks about avoiding car crash by the following vehicle and also offers a mechanism for avoiding accidents at the time of changing lanes. Venkatesh K. [7] has also worked on similar lines and proposed an intelligent system that focusses on avoiding collisions by making use of neural networks and laser sensors. Data collected by the sensor will be prioritized and analyzed for predicting the collision by a two stage prediction system. The proposed system would, then, alert the driver if the possibility of the accident has been established.

Current collision detection system also utilize Lidar sensors [1]. Several works discuss the suitability of Lidar sensors. Frontal Vehicle Collision Warning System [8] highlights that combined use of Radar and Lidar sensors can provide more fruitful results. Works by T.U. Anand Santhosh Kumar, J. Mrudula [6] shed light on yet another important aspect that is usually ignored, i.e., weather conditions. The model proposed by them suggests that ultraviolet rays and infrared rays can perform well in harsh weather conditions where most other existing models would fail.

United States Department of Transportation has also proposed Intersection Collision Avoidance System which utilizes the vehicle to vehicle, dedicated short–range communications (DSRC) to share the safety critical state information.

Many leading automobile companies have equipped their cars with the latest technology for avoiding collisions. Adaptive cruise control and collision avoidance systems are now present in the models of popular car brands but such systems are restricted only to the high end expensive models. Some of them are BMW's Active cruise control and Audi's full speed range active cruise control based on radar which was world's first GPS-guided radar active cruise control.

III. ARCHITECTURAL COMPONENTS OF THE PROPOSED SYSTEM

All the works that have been undertaken till now, focus only on an individual vehicle, i.e, capability of the collision detection system is limited only to avoiding the accident. Such a system does not communicate with other vehicles or objects in its surroundings. This shortcoming of the collision detection system can be overcome by applying IoT techniques to automobiles. An IoT network of connected cars will bear more significance since it would offer Vehicle–to-Vehicle (V2V), Vehicle-to-Infrastructure (V2I) and Infrastructure-to-Vehicle (I2V) communication.

Hence, we propose a model for IoT automobiles and offer insight into the architecture for such a system.

Use of Image Processing for Detecting Objects:

Most of the research done till now implements Lidar, Radar, Ultrasonic sensors for detecting the vehicles with which collision is likely possible. Another possible alternative that can be considered is through the use of cameras. Cameras can detect the vehicle in front and also estimate its size. If the leading vehicle's image is growing constantly at a rate that is greater than the threshold value, an alarm will be triggered to the driver informing him about the likely collision that would occur if appropriate measures are not taken. The growth rate of the image in the camera will be an indicator of the fast speed of the vehicle and hence can be used to avoid collision. The leading car can also be informed of the collision though IoT so that the driver of the leading car can take appropriate measures to avoid the collision. Therefore, IoT will facilitate vehicle to vehicle communication in such cases. Here, thermal cameras and laser cameras can also be used for improving night vision and vision during bad weather conditions when visibility is affected.

CHIPS FACILITATING V2I AND I2V COMMUNICATION:

The term 'Internet of Things' refers to a network of objects where objects communicate with each other. But, most of the times we assume our object to be dynamic, not static. The collision avoidance systems that have been implemented or proposed provide guard against collision from other vehicles. However, a situation may arise when a car collides with a stationary object such as a pole or a cliff in extremely harsh weather conditions due to poor or no visibility. The current collision system would fail in such a case. This loophole can be sealed through the use of IoT.V2I and I2V communication can be done with the help of advanced augmented reality (AR) techniques which provide a real-time direct or indirect view of a physical, real-world environment whose elements are supplemented by computer-generated sensory input which includes sound, video, graphics or GPS data. Implanting sensory chips on stationary objects that make a buzzing sound on sensing an approaching vehicle can be helpful in avoiding collisions and also grease the wheels for V2I and I2V communication.

ZIGBEE FOR ESTABLISHING MESH NETWORK FOR CONNECTED VEHICLES:

In today's world, power consumption is an important factor that can impede an innovation. Same thing is applicable for IoT vehicles. Power crunch and battery drainage can hamper the connectivity among vehicles. Wi-Fi and Bluetooth are the most widely used wireless protocols. However, their drawback is that both require lot of energy to operate and many vehicles in the Internet of the Things model only require a small amount of data to be exchanged. In other words, this is overkill for many applications. Hence, we either need to switch to another low power consuming wireless protocol or we need to use superior quality batteries in order to transform the concept of IoT collision avoidance system into a reality. A potential alternative that we consider appreciable is ZigBee. It's suitability for mesh networks and lower power consumption make it a suitable candidate for IoT. The ZigBee standard allows low-powered devices to transmit information along a network, with each device capable of relaying the data toward its intended destination. If the standards for ZigBee are improved, this would prove to be a good choice for IoT.

INERTIAL NAVIGATION SYSTEM:

Often in some situations, a GPS may not work properly due to weather and signal constraints. In such cases, collision systems that track objects through GPS and GSM technology are bound to fail and would not give correct results. Therefore, an efficient collision system should have some architectural components that provide a backup in case the GPS does not function. Here, an inertial navigation system can compensate. The inertial navigation system makes use of a computer, accelerometers and gyroscopes to compute the position, orientation, and velocity of a dynamic object without the need for external references. The system incorporates dead reckoning to calculate the above mentioned aspects. Inertial measurement units (IMUs) consists of three orthogonal rate-gyroscopes and three orthogonal accelerometers which measure angular velocity and linear acceleration respectively. By processing signals from these devices it is possible to track the position and orientation of a moving vehicle. The above mentioned mechanism clearly illustrates the benefits that an inertial navigation system has over GPS. Consequently, by using inertial navigation system coupled with a receiver, the need for a GPS can be eliminated.

IV. CHALLENGES

After having proposed the key architectural components of an IoT collision detection system, we now want to shed light on a few hurdles that still need to be handled properly in order to ensure an efficient system of connected vehicles.

An important issue that researchers in the field of IoT are still struggling with is security and privacy.

IoT does provide a connected network of vehicles but in V2V communication exposes the vehicle owner to risks that compromise on his safety. We need to ascertain the level to which such compromises can be admissible. The benefits of V2V communication should not be at the cost of endangering our safety.

Developers designing algorithms for IoT vehicles need to consider all possible use cases for sealing all loopholes in the programs. This demands extensive research, diligence and perseverance on the developers' part because a single overlooked use case can be fatal. Tech giants proposing algorithms for intelligent vehicles restrict themselves to only providing an outlying algorithm and are not venturing into developing an executable program code for the same. Hence, the cumbersome task of developing such efficient and effective programs rests on the automobile companies. While developing algorithms, it also needs to be considered that technologies and programming models employed for such purposes are interoperable and backward compatible.

On similar lines, to ensure effective communication among vehicles, a proper routing mechanism needs to be implemented which would offer alternative paths for transmitting data in case a node fails. Also, the transit time of data should be minimal to ensure collision avoidance.

The user interface of the IoT vehicles should be as simple as possible so that it does not make the process abstruse for the common man. Another concern that is of paramount importance is the generation of enormous amount of data when V2V, I2V and V2I communication occurs. The data needs to be managed, organised and filtered so as to retain and transmit only the crucial information. Mechanism to delete the unwanted information should also be formulated to prevent unnecessary aggregation of data. Unwanted data will occupy additional space in servers creating the need for more and more servers.

The above mentioned points clearly indicate that though IoT Collision System is promising, there still exist a number of aspects that need thorough research. While looking into the next stages of the IoT and its technologies there are many opportunities and challenges to face including privacy concerns, security, costs, standards and regulations. Handling all of these myriad of problems is a daunting task and research communities are working for better solutions.

V. CONCLUSION

The above proposed IoT collision detection system aims to provide a solution for reducing the number of accidents by harnessing the concept of IoT.V2V communication can warn the surrounding vehicles of a collision that is likely to occur and hence allow them to take the required action in time. The architectural components of such a system suggested above may prove beneficial and economical. The sophisticated multitude of sensors and chips can help record real time information, which can assist the driver. It proactively estimates collision threats that can be foreseen in the individual's path while driving. Through this paper we have tried to present a model for collision detection with the use of IoT technology that helps in sharing information between different automobiles on the road and provide the well optimized solution thus reducing the risk for the drivers on the road. We also aim to take this research further and examine the deeper nitty-gritties that the proposed system may have and also work upon the refinement of the system.

REFERENCES

[1] Dmitry Namiot, Manfred Sneps-Sneppe, *On IoT Programming*, International Journal of Open Information Technologies ISSN: 2307-8162 vol. 2, no. 10, 2014, pp 25-28.

[2] Jeich Mar, Senior Member, IEEE, and Hung-Ta Lin, *The Car-Following and Lane-Changing Collision Prevention System Based on the Cascaded Fuzzy Inference System*, IEEE transactions on vehicular technology, vol. 54, no. 3, May 2005.

[3] Ovidiu Vermesan, Peter Friess, *Internet of Things-From Research and Innovation to Market Deployment*, Alborg, Denmark, River Publishers, 2014.

[4] R.Sivakumar, Dr.H.Mangalam, *RADAR Based Vehicle Collision Avoidance System used in Four Wheeler Automobile Segments*,

International Journal of Scientific & Engineering Research, Volume 5, Issue 1, January-2014, pp. 763-770.

[5] Stan Schneider. (Oct 9, 2013), *Understanding The Protocols Behind The Internet Of Things.* Available:http://electronicdesign.com/iot/understanding-protocols-behind-internet-things

[6] T.U. Anand, Santhosh Kumar, J. Mrudula. *Advanced Accident Avoidance System for Automobiles.* International Journal of Computer Trends and Technology (IJCTT) – volume 6 number 2–Dec 2013.

[7] Venkatesh, K. *NEFCOP: A Neuro-Fuzzy Vehicle Collision Prediction System,* International Conference on Intelligent Agents, Web Technologies and Internet Commerce, Nov.2005, pp.28-30.

[8] Yimin Wei, Huadong Meng, Hao Zhang, and Xiqin Wang, *Vehicle Frontal Collision Warning System based on Improved Target Tracking and Threat Assessment,* Proceedings of the 2007 IEEE.

Overview of Holographic security

Aman, Riya Pant, Dheeraj, Devanshi, Sandeep Sharma

Department of Computer Science Engineering, DIT University, Dehradun, Uttarakhand, 248009, India

e-mail:amanadhikari2@gmail.com,riyapant.rp@gmail.com,devanshi0114@gmail.com,yadavdheeraj@gmail.com, tek.learn@gmail.com

Abstract: Holography is the new way that changes the way of conventional security systems 180 degrees. Gaining tremendous attention in various fields, holography puts a firm foot in security. Holograms are very difficult to forge and thereby offer more efficient security. Its efficiency has found itself convenient in banknotes around the world. Our paper emphasizes on the key security measures implying holography and its advantages over others.

I. INTRODUCTION

Holograms are basically photographic recording of a light field, and is used to display a full three-dimensional image of a object, which is visible even without the aid of special glasses or other optical instruments. Holography was a method accidently developed by Dennis Gabor in 1947,in an effort to improve electron microscopes. Holography is a technique that enables a light field, which is generally the product of a light source scattered off objects, to be recorded and later reconstructed when the original light field is no longer resent. During the recording process a standing wave interference pattern is generated where the reference beam and object beam interfere within the recording layer.

Holography

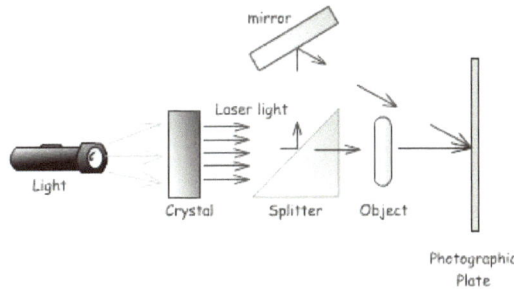

Fig [1]

When the two laser beams reach the recording medium, their light waves intersect and interfere with each other. It is this interference pattern that is imprinted on the recording medium and with same technique messages can be encrypted on photographic plates. The interference pattern can be considered an encoded version of the scene, requiring a particular key — the original light source — in order to view its contents. This missing key is provided later by shining a laser, identical to the one used to record the hologram, onto the developed film. When this beam illuminates the hologram, it is diffracted by the hologram's surface pattern. This produces a light field identical to the one originally produced by the scene and scattered onto the hologram and thus decrypting the message. Thus information can be secured via holograms. Holograms have widened the doors to optical security systems.

II. OPTICAL SECURITY SYSTEMS

The advantage of the optical method in a security system is the fast processing for decoding an encrypted image and identifying it. In practical optical security systems, a digital technique may be used for the image encryption, since the time required to calculate an encryption pattern is not essential and the encryption of an image and printing the encrypted hologram on a card may be done by offline. On the other hand, the fast processing is required for decoding an encrypted image and verifying it. Accordingly, optical technique is very suited for such processing.

The optical security system consists of three parts.(a) Image encryption, (b) Image decryption, and (c) identification systems. Fig. 2(a) is an encryption system of a target image. An image, for example a finger print image, is encrypted with an encoding key with holographic technique and the encoded image is binarized according to a certain rule to make it easier for reading the image and to match the post processing system. The encrypted binary image printed on a security card is optically decrypted with the decoding key in Fig. 2. The decoded image is compared with a test image for identification with optical joint transform correlation as shown in Fig. 2.

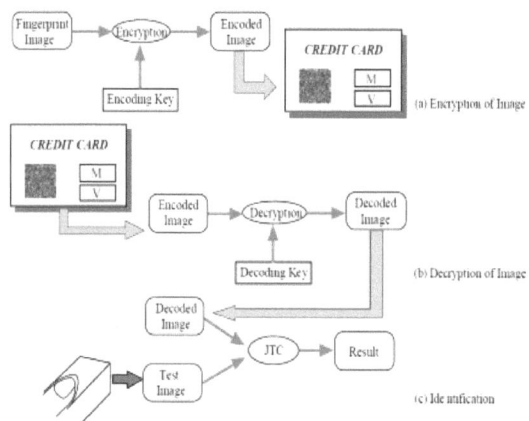

Fig [2]

III. ENCRYPTION AND DECRYPTION OF HOLOGRAM

In the image encryption process, an image with a random phase mask is jointly Fourier transformed with another random pattern, which is a key for encryption and decryption a hologram is formed at the Fourier plane. The holographic fringe terms are given by

$$H\ (u,v)\ =\ F(u,v)G^*\ (u,v)\exp(-i4\pi dv)\ +\ F^*(u,v)G(u,v)\exp(i4\pi dv)\ -(1)$$

where F(u,v) and G(u,v) are the Fourier transformed functions of the image with a random phase, f(x,y), and the random pattern g(x,y) used as an encryption key, respectively, 2d is the separation between the centers of two functions, * denotes the complex conjugate, and H(u,v) is the resultant hologram. To reconstruct the image from the encrypted pattern, the hologram is illuminated by the same random phase pattern used for the encryption as shown in Fig. 4. For the decryption corresponding to Eq.(1), we obtain

$$p(x,y)=f(x,y-d)\otimes g(-x,-y)\otimes g(x,y)+f(-x,-y)\otimes g(x,y)\otimes g(x,y)\otimes \delta(x,y+3d)$$

where $\otimes$ denotes the convolution operation. The second term is the convolution between the image and the random function and it is a noise term in the reconstruction. Two terms can be spatially separated with each other. Thus, the encrypted image is successfully decrypted in the image plane without noise terms.

IV. RECONSTRUCTION OF BINARY HOLOGRAM

In actual applications, such as in credit card identification, a binary hologram is suited for acquiring electronic image and digital-electronic pre-processing. Thus, the use of the binarization of hologram is one of excellent methods to make a robust security system, so that we employed a binary hologram as an encrypted pattern in the following. The calculated hologram is binarized according to the sign of each element of the composite pattern.

 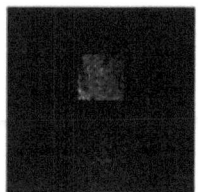

Fig [3a]   Fig[3b]

Fig. 3 is the result of the encrypted binary hologram. In the figure, the hologram that has a (0,1) binary distribution is printed on a card as a black and white pattern. However, for the reconstruction of the binary hologram, the value of each pixel is assigned to +1 (exp(i0)) when the pixel hologram has a value of 1, while it is -1 (exp(iπ)) for 0. A hologram that has (0,π) phase distribution can be easily realized by using a phase modulation spatial light modulator such as a parallel aligned liquid crystal display.

The hologram that has 0 and π phase distribution has the advantage in the reconstruction, since the zero-th order diffraction is eliminated in the reconstruction pattern. Fig. 3(b) is the decrypted pattern. The hologram and the reconstructed image have the size of 256×256 pixels. The binarization of hologram is suited for printing it on a credit card in practical use.

V. USES OF HOLOGRAM

1.FMCG(fast moving consumer goods) are products that are sold quickly and at relatively low cost.

2. Hologram dockets for vehicle number plate Some vehicle number plates on bikes or cars have registered hologram stickers which indicate authenticity. For the purpose of identification they have unique ID numbers.

3. High security holograms for credit cards These are holograms with high security features like micro texts, nano texts, complex images, logos and a multitude of other features. Holograms once affixed on Debit cards/passports cannot be removed easily. They offer an individual identity to a brand along with its protection.

REFERENCES

[1] fig 1- https://lifeos.wordpress.com/2008/05/23/holographic-properties/

[2] fig 2,fig 3-] Optimization of Hologram for Security Applications Junji Ohtsubo

[3] Javidi B.; & Ahouzi E. (1998). Optical Security System with Fourier Plane Encoding, *Appl.*

[4] Optimization of Hologram for Security Applications Junji Ohtsub

Investigation of x-ray spectra in low energy ion-atom collisions

Tulika Sharma, Avnee Chauhan
Department of Physics
Amity institute of applied Sciences
Amity University
Noida, India.
sharma.tulika8936@gmail.com

Punita Verma
Department of physics
Kalindi College
University of Delhi
Delhi, India
drpunitaverma.nature@gmail.com

Abstract— Investigation of ion-atom collision phenomena have been initiated at the recently developed experimental set up of 75° beam line of Low Energy Ion Beam Facility (LEIBF) at Inter University accelerator Centre (IUAC), New Delhi. The LEIBF facility with its 10GHz ECR-ion source provides ion beams of low energy ranging from 50 keV to 3.0 MeV of few hundred nA current. The purpose of this experiment was to understand the physics behind the ion-atom collision of slow highly charged ions with a thick target as reported by Ying et al. [1] and others. The atomic collisions at low energies find wide ranging applications in astrophysical plasmas, in planetary atmospheres and they play a major role in energy transfer as well as charge exchange processes. Xe^{q+} ions ($8<q<13$) of 2 to 3 MeV were bombarded on solid state targets of Au and Zr having a thickness of 640 mg/cm^2 and 250 mg /cm^2 respectively (latter on a carbon backing of 40 mg/cm^2). Projectile as well as target X-rays were observed using KETEK's SDD and a Canberra LeGe X-ray detector. Effects of energy and charge state variation on target and projectile x-rays were investigated.

Keywords—ion-atom collision; detector; data analysis; LEIBF; ECR ion source.

I. INTRODUCTION

X-ray emission was first observed by ROENTGEN in 1895. The behavior of X-ray was then considered to be mysterious and was named so. It has long been classified as electromagnetic radiation that originates from an atom, and is known to be produced during an ion-atom collision. The reason that X-ray study is a major interest of research is because of the complexity introduced by using heavy ions as projectiles. The main reason for the production of X-rays is the creation of vacancy in an electronic shell of an atom due to the incident ions. Whenever an electron from outer shell fills this vacancy, there is emission of an X-ray. Fig.1. shows the schematic diagram of energy levels for an atomic electron and the consequent x-ray transitions. The atomic collisions at low energies find wide ranging applications in astrophysical plasmas, in planetary atmospheres and they play a major role in energy transfer as well as charge exchange processes. The purpose of this experiment was to understand the physics behind the ion-atom collision in slow highly charged ions interacting with the thick target as reported by Ying et al. [1] and others.

Xe^{q+} ions ($8<q<13$) of 2 to 3 MeV were bombarded on solid state targets of Au and Zr having a thickness of 640 mg/cm^2 and 250 mg /cm^2 respectively (latter on a carbon backing of 40 mg/cm^2). Projectile as well as target X-rays were observed using KETEK's SDD and a Canberra LeGe x-ray detector which were mounted at 90° and 45^0 to the beam line in the experimental chamber respectively. The data acquisition was done using software CANDLE. Analysis of ion-atom data was done using two different softwares namely ORIGIN[2] and MAGIC PLOT[3]. Effects of energy and charge state variation on the x-ray spectra were investigated.

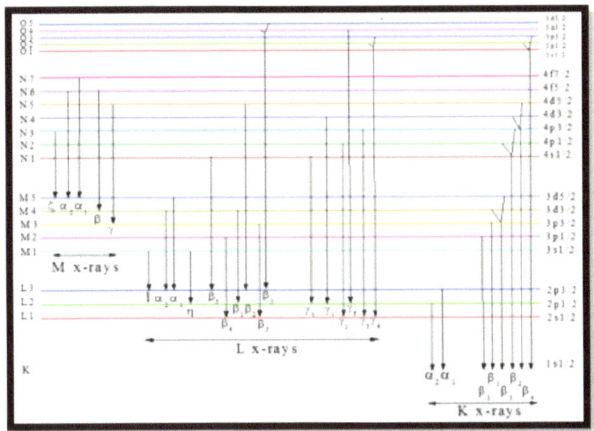

Fig.1: Schematic diagram of energy levels for an atomic electron and x-ray transitions associated with it.

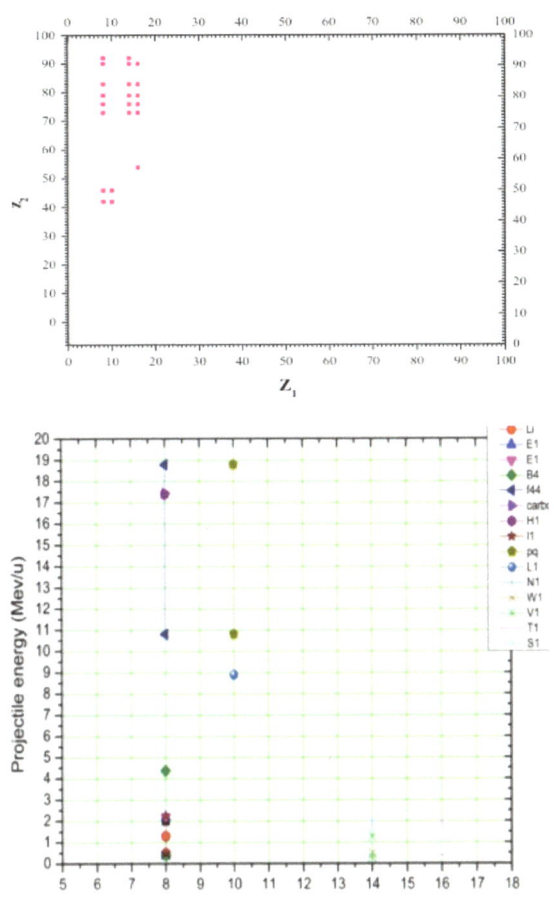

Fig.2a. Graphical representation of literature survey of multiple ionization investigations by other authors for target atomic no. Z_2 v/s projectile atomic no. Z_1.

A significant shift of X-ray lines towards higher

energies has been observed by other authors when the targets are bombarded with heavy ions, this effect observed was due to multiple ionization. An extensive literature survey was conducted regarding the various earlier investigations of multiple ionization [4-13]. The survey is represented in graphical form in Fig.2. It was inferred that the phenomenon of multiple ionization has been studied for mostly heavy and highly energetic ion beams like Oxygen, Sulphur and Silicon from 6 MeV to 20 MeV.

II. EXPERIMENTAL DETAILS

LEIBF FACILITY @ IUAC

The low energy ion beam facility (LEIBF) at IUAC provides three different beam lines of low energy belonging to the fields of Atomic physics, Molecular physics and Material Science. The ion-atom collision data in this experiment was recorded at the atomic beam line at an angle of 75°. Fig.3. shows the schematic diagram of three different beam lines. Here we will discuss about the 75° beam line which is used for conducting atomic physics experiments.

ECR ION SOURCE AT IUAC

The ECR ion source makes use of the electron cyclotron resonance to ionize plasma. Microwaves are injected into a volume at the frequency corresponding to the electron cyclotron resonance, defined by the magnetic field applied to a region inside the volume. The volume contains a low pressure gas. The alternating electric field of the microwaves is set to be synchronous with the gyration period of the free electrons of the gas, and increases their perpendicular kinetic energy. Subsequently, when the energized free electrons collide with the gas in the volume they can cause ionization if their kinetic energy is larger than the ionization energy of the atoms or molecules. The ions produced correspond to the gas type used, which may be pure, a compound, or vapor of a solid or liquid material. ECR ion sources are able to produce singly charged ions with high intensities. For multiply charged ions, the ECR ion source has the advantages that it is able to confine the ions for long enough for multiple collisions and the low gas pressure in the source avoids recombination.

ECR ion source at IUAC is installed on a high voltage deck. This facility is based on permanent magnet which is designed to confine the plasma radially and axially. This facility provides opportunity for research in field of atomic physics, surface physics and material science.

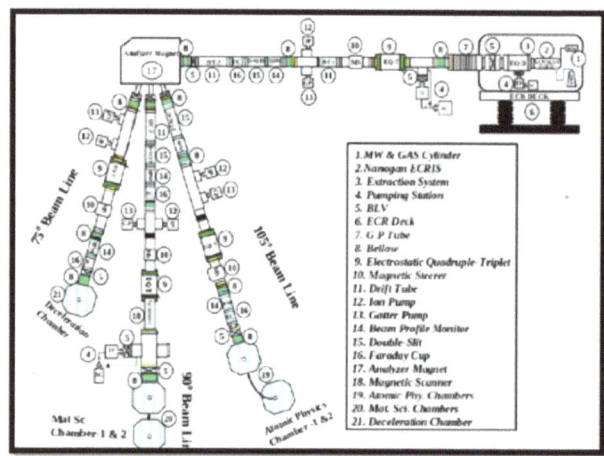

Fig. 3 Schematic diagram of LEIBF at IUAC

Detectors

KETEK SDD and LEGe detectors were used to measure the X-ray radiations. The characteristics of SDD detectors are high count rates, comparatively high energy resolution and peltier cooling. The material used in SDD detector is highly pure silicon crystal with a very low leakage current. Due to high purity of material, Peltier cooling can be used instead of traditional liquid nitrogen. The detection of the signal in SDD is based on the principle of ionization. It measures the energy of photon by the amount of ionization it produces in the detector material. The resolution of the KETEK SDD detector used in the experiment is 125eV at 5.9keV of Fe Ka.

LEGe are the low energy germanium detectors which are mostly used to study gamma-ray spectroscopy and X-ray spectroscopy at low energies. The LEGe detector also measures the energy of a photon. The drawback of germanium detector is that it must be cooled to liquid nitrogen temperature to produce spectroscopic data. At high temperature the electrons may jump over the band

gap and reach the conduction band due to thermal agitation. The resolution of the KETEK SDD detector used in the experiment is 145eV at 5.9keV of Fe Ka.

III. EXPERIMENTAL CHAMBER

The experimental chamber used in the experiment is shown in Fig.4. The chamber consists of several ports. Beam enters through one of the ports as shown. The LEGe detector is kept at an angle of 90° from the beam entrance port. The KETEK SDD detector is placed at an angle of -45° from the beam entry port. Target ladder is inclined at 45° to the beam line. Other things like camera and light bulbs

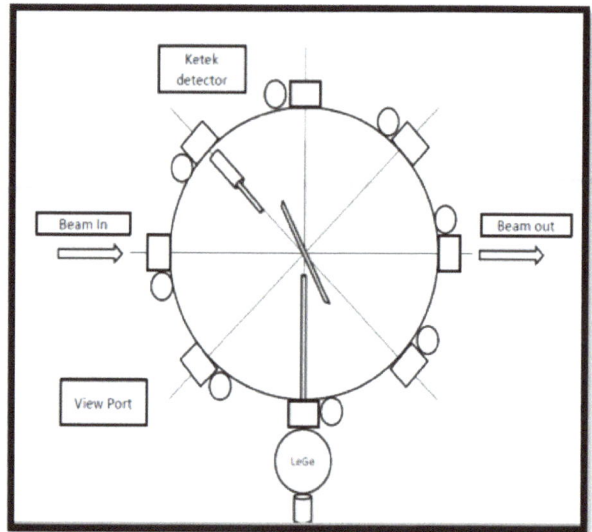

Fig.4. Schematic diagram of experimental chamber

can also be placed according to the requirement.

Vacuum of 3×10^{-3} mbar is created in the chamber, obtained using rotary pump and turbo pump. Rotary pump provides the rough vacuum of the order of 10^{-3} mbar. The fine vacuum of 3×10^{-3} mbar is obtained using the turbo molecular pump.

IV. ENERGY CALLIBRATION AND EFFECIENCY MEASUREMENT OF KETEK SDD DETECTOR

Proper energy calibration of the detector is essential for accurate measurement of X-ray cross-sections in an ion-atom collision experiment; hence we have measured the energy calibration of KETEK detector using radioactive sources of known energies. Accurate calibration should involve a standard source with X-ray energies that are not widely different from those to be measured in the unknown spectrum. Radioactive source of ^{241}Am was used for the energy calibration of SDD detector.

Once energy calibration has been established over the entire energy range of interest, a calibration curve relating x-ray energy to channel number is normally derived. Common techniques involve the least square fitting of a polynomial of the order one, generally of a straight line equation, where energy is on Y axis in keV and channel number is on X axis.

ELECTRONIC COMPONENTS USED FOR CALIBRATION

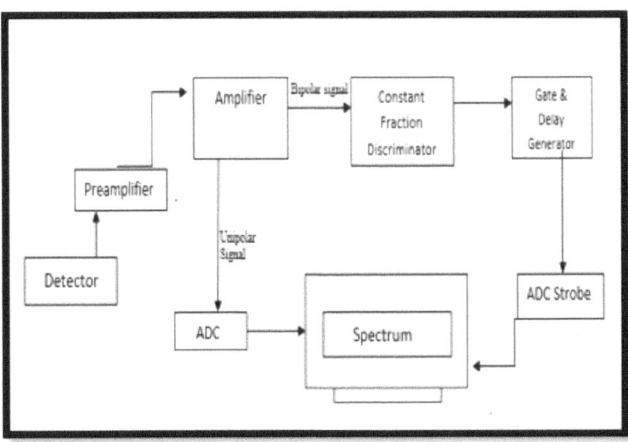

Fig.5. Schematic electronics diagram for energy calibration measurement with KETEK detector

•Preamplifier - The function of preamplifier is to amplify a low level signal to high level. A list of common low level signal sources include pick up, microphone, and transducer.

•Amplifier - Amplifier uses a small input signal to create a large output signal while minimizing distortion of the waveform. Amplifiers are used for both electrical and optical signals.

•Constant Fraction Discriminator - It is an electronic signal processing device, designed to mimic the mathematical operation of finding a maximum of a pulse by finding the zero of its slope. Some signals do not have a sharp maximum, but have short rise times.

•Timing Filter - It is designed to meet the needs of most timing experiments and provide good timing resolution over wide range of pulse amplitudes.

•Analog to Digital Converter (ADC) - It converts the analog signal into digital signal.

Fig. 5. shows the schematic diagram of the electronics set up for such a measurement with KETEK detector.

The radioactive source of ^{241}Am.was placed in front of the detector and the X-ray spectra were recorded by the SDD detector. With the prior knowledge of the energy of the transitions occurring in these radioactive sources, the energy calibration of the detector was done by plotting the channel number corresponding to the energy of the X-ray lines emitted by the radioactive sources. A first order polynomial was fitted to the plotted curve. This polynomial could be used for identification of the characteristic X-rays of a target element and the presence of any impurities in the target. The spectra of radioactive source recorded by KETEK SDD detector has been analysed and the calibration curve for amplifier gain 100 and CFD threshold -0.857 mv is shown in Fig. 6.

DATA ANALYSIS

Xe^{q+} ions ($8<q<13$) of 2 to 3 MeV were bombarded on solid state targets of Au and Zr having a thickness of 640 mg/cm^2 and 250 mg/cm^2 respectively (latter on a carbon backing of 40 mg/cm^2). Typical M X-Ray spectra of Au and L X-rays spectra of Xe and Zr are shown in Fig.7 and 8. The deconvolution of spectra into different M x-ray lines for Au and L x-ray lines for Zr was done with

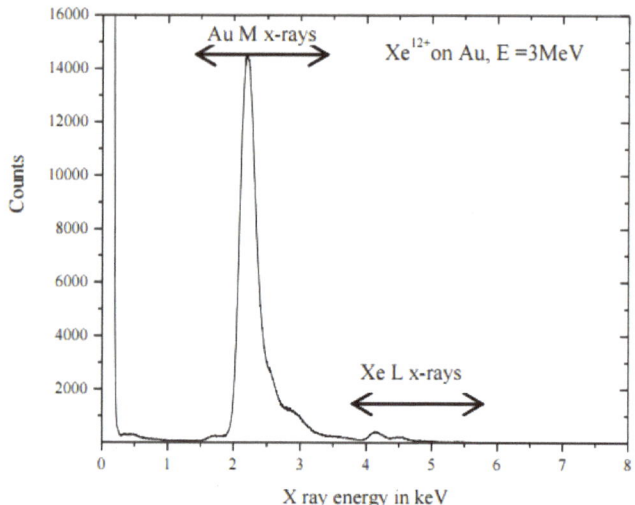

Fig.7. Spectra of 3 MeV Xe^{12+} incident on Au and Zr targets

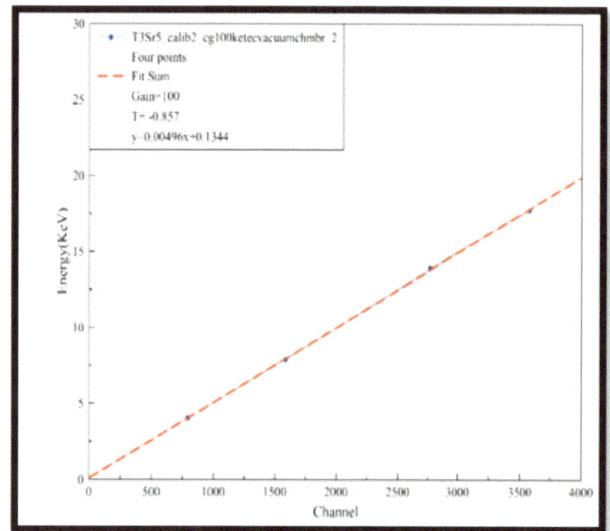

Fig.6. Energy calibration of KETEK SDD detector with Am^{241} source

gauss fitting method using the program MAGIC PLOT. It was important to subtract the background area as improper subtraction may lead to uncertainties in the cross section calculations. Gauss fitting curve of Au spectra is shown in Fig. 8. A shift in the energy of x-ray transitions is observed as compared to the standard x-ray energies by Bearden et al. [4]. This effect was observed due to the phenomenon of multiple ionization.

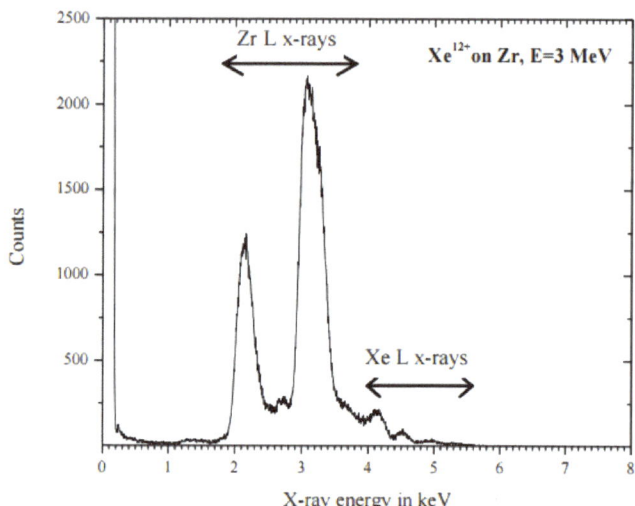

Fig.8. Spectra of 3 MeV Xe^{12+} incident on Au and Zr targets

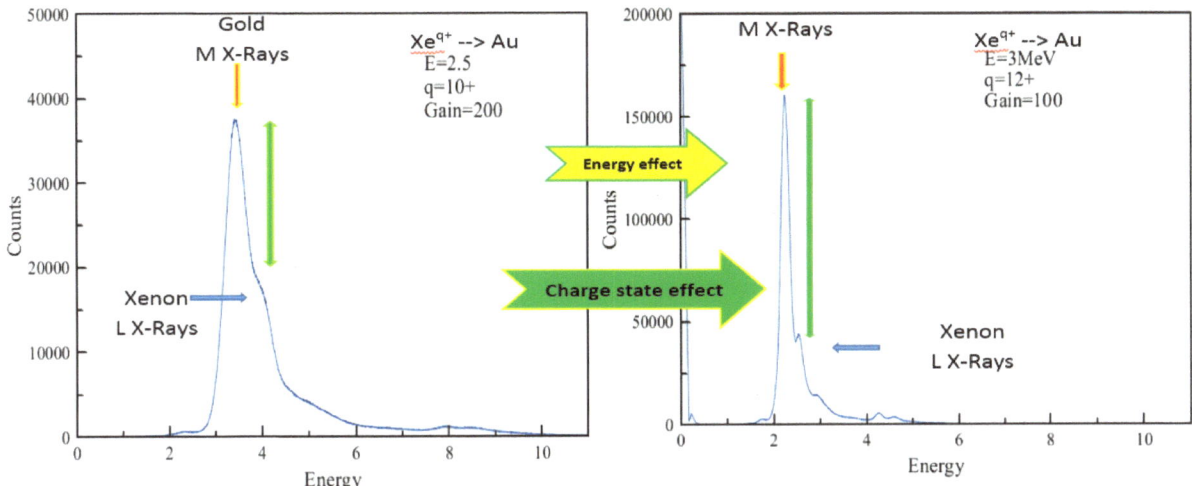

Fig.9. Gaussian peak fitting of Au M X-rays

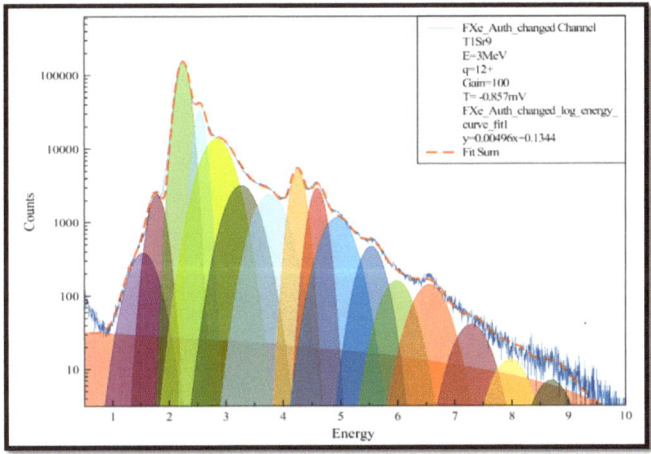

spectra.

Fig.10. Comparison of two spectra of gold bombarded by $Xe^{10+ \& 12+}$ ions

V. CONCLUSION

From a literature survey of earlier investigations on multiple ionization (Fig.2a,b), it is concluded that the phenomenon of multiple ionization has been studied mostly for heavy, highly energetic ion-beams like oxygen, sulphur and silicon ranging from 6 MeV to 20 MeV. The targets used were of high atomic number Z_2 (>46). During multiple ionisation, It was observed that when the same target is bombarded with the projectiles of different energy and different charge states, a significant change in the x-ray peak energies was seen towards

higher values. Till date the dynamics processes of atomic collisions are not completely understood at the low energy regime and hence a lot of experiments are still possible with different combinations of projectile beams and targets.

It was also noted that the energy of the peaks which were found by gauss fitting did not match the energies of single ionization which leads to the fact that the phenomenon of multiple ionization took place and hence a shift was observed in the energy.

The projectile charge state effect on target x-ray spectra (Fig.10) leads to an increase in the relative x-ray peak intensity ratios. It was observed that when the target is bombarded with the projectiles of different energies, marked change in the x-ray spectra occurs. Fig.10. shows the proper comparison of the effects observed.

ACKNOWLEDGEMENT

One of the authors, PV is thankful to Inter University Accelerator Centre (IUAC), New Delhi for providing the beam time at LEIBF. Thanks are due to Mr. Kedar mal for providing the beam and to C.P. Safvan and S. Kumar for constant help during beam time. The experiment would not have been possible without their help.

REFERENCES

[1]   Ying et al. Sci China Ser G-Phys Mech Astron | Sep. 2008 | vol. 51 | no. 9 | 1240-1244

[2]   http://magicplot.com/downloads.php

[3]   http://www.originlab.com/

[4]   Bearden et al. Rev. Mod. Phys. 39, (1967) 78-124

[5]   Banas et al. NIM B154, (1999) 247–251

[6] Banas et al. NIM B 195, (2002) 233-246

[7] Banas et al. J Phy B 35, (2002) 3377-3598

[8] Banas et al. Phys. Rev. A 68 (2003) , 022705

[9] Czarnota et al. NIM B 201, (2003) 133–138

[10] Czaronata, et al. Braz. J. Phys.36 (2006)

[11] Czaronata, et al. J Phys. 58 (2007)

[12] Fijal Banas et al. Phys. Rev.A 77, (2008) 032706

[13] Czarnota et al. Phys. Rev. A 79 (2009) , 032710

Simulations for designing a light rotor wind turbine for Delhi Metro

P. Verma, D. Bhardwaj, N. Sharma, T. Sharma, K. Kumari, R. Gupta, A. Fanda, M. Sachdeva,

R. Reghu, S. Tanwar, S. Chaudhary, P. Jha.

Department of Physics, Kalindi College, University of Delhi, New Delhi, India. drpunitaverma.nature@gmail.com

Abstract— **All over the world people are shifting towards 'clean and green' renewable sources of energy to reduce the dependence on non-renewable sources for electricity generation. Wind energy is one such renewable resource which can be used as an alternate for fossil fuels. This project is an effort in the direction of providing alternative energy source by utilizing the wind energy produced by fast moving underground metros. The idea can be realized by installation of small wind turbines on Delhi metro platforms or inside the tunnels of underground metros. To select appropriate shape and size of turbines, we have to conduct the simulations to estimate the amount of power that can be produced through this setup. To accomplish this, many simulation softwares like PROPID, Warlock, SOWFA, etc. were investigated and finally Warlock software was selected due to ease of use, availability and better results. For a proper simulation of wind velocities, knowledge of complete wind profile of platforms of different Delhi metro stations was necessary. Hence wind velocities were measured with the usage of anemometer at 66 metro stations. It was then concluded that high velocities were observed in underground metro stations only, as in elevated stations wind spreads all over due to open area. After this, detailed survey was conducted at few underground stations and it was concluded that higher values were observed at entry/exit of metro.**

Keywords— *simulation software, wind energy, anemometer, metro.*

I. INTRODUCTION

Depletion of non-renewable resource has shifted the interest of people towards utilization of renewable resources for generation of power. Undoubtedly wind energy is one of the cleanest renewable energy source for production of power. In the last few decade, a growing interest in renewable energy resources has been observed. Unlike other renewable energy

sources, wind energy has become modest with conventional power generation sources and this has led to the growth of wind turbine industry as well. Unfortunately wind energy is not available in metropolitan cities in ample amount but underground metro stations are a possibility. Here, one can experience a huge gush of wind while standing on platforms. This wind current arises due to the pressure difference at the ends of the tunnel because of the high speed metro trains. The resulting wind has a high kinetic energy and can be converted into an appreciable amount of electrical energy through a wind turbine[1,2]. Wind turbines work by converting the kinetic energy of the wind into rotational energy and then into electrical energy.

To extract maximum kinetic energy from the wind an appropriate turbine has to be used designed to give maximum possible efficiency [2]. A wind turbine should always extract the highest possible power from the wind and energy generation should be more than the energy loss due to air drag. Hence it is important to design an appropriate turbine taking into account all the factors.

II. METHODOLOGY

A short review to our Initial Steps

Our survey of metro platforms and tracks gave us some positions for the installation of turbines. A three bladed HAWT (Horizontal Axis Wind Turbine) with blade length of 30 cm was installed at one of the sites given by DMRC (Fig.1). Unfortunately, the location was not appropriate to get even threshold velocity as there was a wall at the back side which obstructed the free flow of wind produced. Another reason of the failure was that the blade length of the turbine was too small to catch sufficient wind to move its heavy rotor. So, further measurements were made by 5 bladed HAWT having a lighter rotor.

Fig.1. Three bladed turbine at the Chawri Bazar metro station

Outcomes with 5 bladed HAWT

We have tested five bladed turbine with a cut in speed of 3 m/s at different stations and it gave better results with high RPM (Rotation per Minute) and less energy loss.

Survey and preliminary observations showed that the turbine moved at a higher speed when kept at a height of 5 feet from the ground. This observation was made at 3 different metro stations, namely Kashmere gate, Chawri bazaar and Chandni Chowk in December 2014. After theoretical calculations with the wind velocity observed at above mentioned stations it was estimated that 0.15kWh (approximate working of 10 hours a day) of energy can be harnessed with a single 5 bladed HAWT at a wind velocity of approximately 6m/s.

III. DATA COLLECTION (WIND PROFILING)

Fig.2. Five bladed HAWT

The aim of wind profiling was to find the location of maximum wind velocity on the platform.

In December 2015, observations were taken at 66 metro stations (elevated as well as underground)[3]. It was found that elevated platforms show very low values of wind velocities ~0.2 m/s as compared to the underground metro stations ~12 m/s. This is because in elevated stations, the wind velocity reduces due to larger open area while in underground tunnel there is high pressure due to narrow opening.

So, in August 2016, we measured the wind velocities only in underground metro stations. This time we have chosen 6 underground metro stations (5 in yellow line and 1 in violet line for comparison). The readings were taken by a group of 12 B.Sc. students using 4 anemometers. By holding the anemometers at various heights (~90cm/175cm/220cm), velocities were noted at and behind the yellow line at two more points (i.e. at yellow line, 120cm behind the yellow line and 240cm behind the yellow line). Similar readings were taken on alternate floor tiles throughout the length of the platform. Videos of observations were also recorded. The collected data was tabulated and related graphs were plotted (in origin software) for wind Velocity vs. Length of the platform.

Observations of wind profiling at Saket Metro Station

Tile Number	Length along the yellow line (cm)	Velocity on the yellow line (m/s)			Velocity at 120 cm behind the yellow line (m/s)			Velocity 240 cm behind yellow line (m/s)		
		Height (cm)								
		95	175	220	95	175	220	95	175	220
4	240	-	-	6.6	-	-	-	-	-	-
6	360	-	-	7.1	-	-	-	-	-	-
8	480	-	-	7.2	-	-	-	-	-	-
10	600	-	-	7.6	-	-	-	-	-	-
11	660	7.4	9.7	7.1	4.7	5.9	-	-	-	-
13	780	7.7	9.5	7.4	5.3	4.6	-	-	-	-
15	900	8.3	10.4	7.2	4.7	6.2	-	-	-	-
17	1020	8.3	9.5	8.6	4.5	5.9	-	-	-	-
19	1140	8.4	9.2	8.6	5.2	6.9	-	3	4.4	-
21	1260	8.6	7.6	9.2	6.2	6.9	-	3.8	3.9	-
23	1380	8.7	9.4	8.6	6.3	7.6	-	4.6	5.3	-
25	1500	8	8.6	8.6	5.6	6.6	-	3.6	5.1	-
27	1620	7.4	8.3	8.8	5.3	6.9	-	4.9	5.5	-
29	1740	6.8	7.2	8.2	4.6	6.8	-	5.6	5.5	-
31	1860	7.3	8.9	8.5	4.3	7.3	-	6.3	5.9	-
33	1980	6.8	8	7.9	4.1	6.8	-	5.7	5.3	-
35	2100	9.7	9.3	8.3	7.5	8.9	-	5	7.6	-
37	2220	9.9	9.8	9	7.6	8.6	-	5.4	7.4	-
39	2340	9.9	10.1	9.3	7.9	8.8	-	5.6	8.1	-
41	2460	9.3	8.3	9.8	6.1	8.4	-	5.3	6.9	-
43	2580	7.7	9.9	9	7.3	9.1	-	6.5	6.6	-
45	2700	7.5	11.2	9.1	7.5	8.6	-	6.6	6.7	-
47	2820	7.8	10.9	9.1	7.7	10	-	6.7	7.3	-
49	2940	6.9	10.7	6.2	6.7	9	-	5.6	6.9	-
51	3060	6.2	8.0	6.3	6.4	6.6	-	2.5	6.3	-
53	3180	6.2	6.5	5.9	6.2	6.6	-	3.7	6.0	-
55	3240	6.6	8.2	5.5	6.4	7.4	-	3.5	7.0	-
57	3420	6.1	7.4	8.3	6.0	6.1	-	3.2	6.9	-
59	3540	7.6	4.0	8.4	7.3	8.4	-	5.2	6.6	-
61	3660	7.7	3.3	8.	7.	8.	-	4.	5.	-

		4		3	3	1		9	2				1		4				6	
63	3780	7.7	8.9	7.7	7.1	8.5	-	5.6	5.9	-215	12900	1.7	2.2	-	1.4	-	0.9	-	-	2.3
65	3900	7.1	3.1	5.6	6.7	7.4	-	5.2	4.9	-225	13500	1.2	2.0	-	1.7	-	0.8	-	-	2.8
67	4020	3.0	2.6	6.2	6.9	7.0	-	3.9	1.3	-235	14100	3.9	2.1	-	-	-	1.5	-	-	-
69	4140	2.9	2.4	6	6.4	6.7	-	3.3	1.3	-245	14700	3.3	2.0	-	-	-	1.1	-	-	-
71	4260	2.8	2.9	5.6	6.5	7.5	-	3	1.4	-255	15300	2.9	2.0	-	-	2	1.4	-	-	-
73	4380	2.6	2.3	4.0	6.0	7.0	-	2.3	2.6	-265	15900	2.0	2.0	-	-	-	1.3	-	-	-
75	4500	2.8	5.1	4.2	7.3	7.3	-	5.2	5.8	-275	16500	2.2	2.9	-	-	-	1.5	2.1	-	0.2
77	4620	2.9	4.6	3.8	7.0	7.3	-	5.0	6.1	-285	17100	2.0	2.7	-	0.8	-	1.5	0.6	-	0.5
79	4740	2.8	4.9	4.0	7.0	7.9	-	5.3	6.6	-295	17700	2.0	2.4	-	1.3	-	0.6	0.9	-	0.3
81	4860	1.8	4.4	2.1	6.4	6.6	-	4.8	5.0	-305	18300	1.9	2.4	-	1.6	-	2.3	1	-	0.2
83	4980	3.8	2.0	0.6	3.1	2.0	-	2.4	0.6	-										
91	5460	1.7	7.2	-	-	2.1	3.4	-	-	-										
99	5940	0.5	5.8	-	-	2.2	1.9	-	-	-										
107	6420	0.3	5.4	-	-	2.2	1.6	-	-	-										
115	6900	0.4	5.0	-	-	0.5	1.6	-	-	-										
125	7500	2.6	7.7	-	-	3.9	5.5	-	-	-										
135	8100	1.4	7.2	-	-	3.7	4.5	-	-	-										
145	8700	2.6	7.0	-	-	3.7	5.3	-	-	-										
155	9300	2.6	6.3	-	3.1	4.6	-	-	-	-										
165	9900	0.3	8.0	-	0.6	4	-	-	-	-										
175	10500	5.1	7.2	-	0	4	-	-	-	-										
185	11100	5.1	7.0	-	0	3.8	-	0.2	-	3.2										
195	11700	4.8	6.6	-	0	-	-	-	-	-										
205	12300	2	2.7	-	1.	-	0.	-	-	2.										

IV. DATA INTERPRETATION AND RESULTS

After a detailed investigation followings are the observations:

At the starting location on platforms, higher velocities are observed at various heights. Higher values are observed at more heights only (~170cm/220cm).

We get a dip in the velocity in the middle region of the platform as it was wider, open and more crowded area.

High velocities are also observed when the two trains on opposite platforms come together. This is due to increase in wind pressure from both the sides.

It is also observed that the anemometer shows high values even before the metro train actually enters the platform i.e. when it is somewhere in the tunnel.

The wind velocity is more when train is in the tunnel than when it enters the platform. This is because in tunnel, the air pressure is high and when it enters the low pressure area (platform), air spreads all over resulting in lower value of wind velocity.

Racecourse, AIIMS and Saket metro stations gave high velocities in the range of 8.0m/s to 11.2 m/s approximately.

Plot of wind velocity along the length of the platform at Saket metro station has been shown in Fig.3.

V. COMPUTER SIMULATION BY SOFTWARE ANYLOGIC [4]

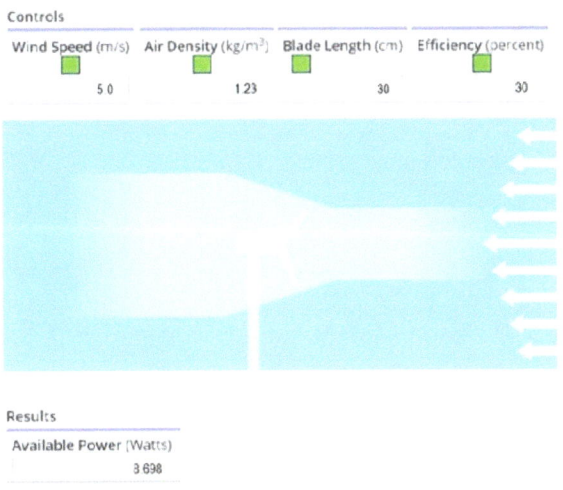

Earlier we had done simulations using

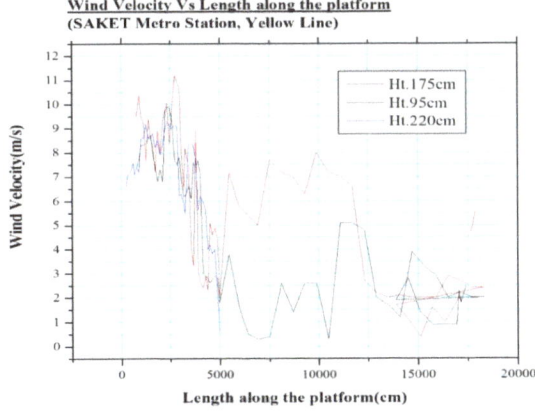

Fig.3. Plot of Wind Velocity Vs Length along the platform at Saket Metro Station,

software ANYLOGIC to get a preliminary idea of what parameters may affect the working of a wind turbine.

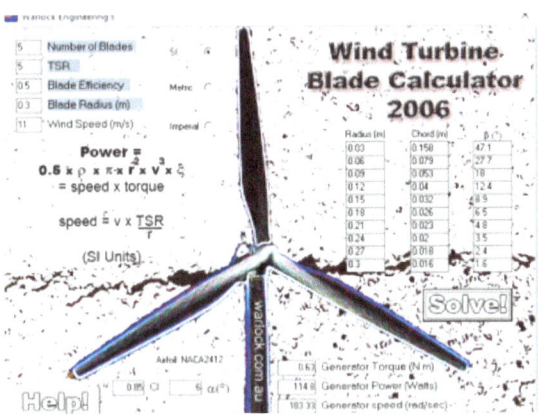

Fig.4. No. of blades=5, TSR=5, B.E =0.5, B.R=0.3, W.S= 11, PG=114

Computer simulations:

After Data analysis we did some simulations to get an idea of power production that we can get from the above data. We have studied various simulation softwares like PROPID[5], SOWFA[6], Warlock [7]etc. In PROPID, they allow us to specify directly the peak power for a stall-regulated wind turbine. After studying these simulation software we have chosen ANYLOGIC and Warlock software to do the simulations due to ease of use, availability and better results.

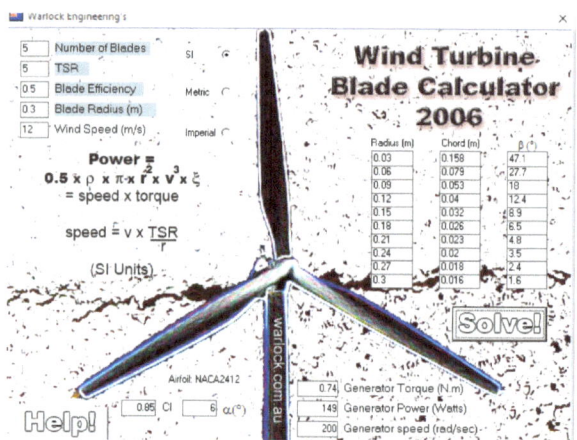

Fig.5. No. of blades=5, TSR=5, B.E =0.5, B.R=0.3, W.S= 12, PG=149

Parameters required by Warlock software:

Number of blades

TSR (Tip Speed Ratio)

Blade efficiency

Blade radius

Wind velocity

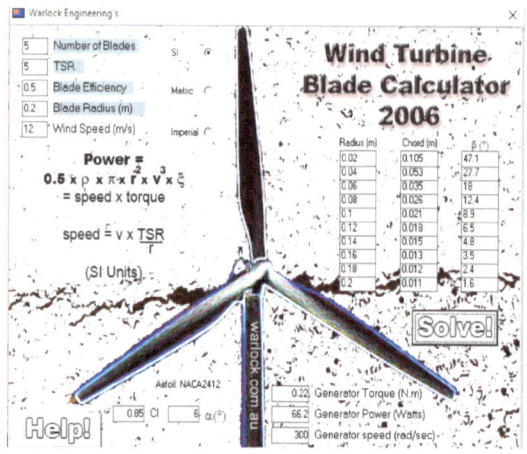

Fig,6. No. of blades=3, TSR=6, B.E =0.5, B.R=0.2, W.S= 12, PG=66.2

Simulations were performed for 3 and 5 number of blades; Blade Radii (BR) = 0.1, 0.2, 0.3; Blade Efficiency (BE) = 0.2, 0.3, 0.4, 0.5; Tip Speed Ratio (TSR)= 4 and 5. This software gave us the values of Generator power (PG) in watts, Generator Torque (Newton meter) and Generator speed (m/s). Following are some of the warlock simulations performed:

The power generated simulated for this software has been plotted in Fig. 7 for the particular case of no = 5, TSR = 5, Blade efficiency= 0.4, Blade Radius = 0.2 m

VI. CONCLUSION

From the detailed survey of wind profiling at different underground metro stations it was found that wind velocity at saket metro station reached as high as 11m/s. Moreover, it could be learnt from the simulations that the maximum power of 149 W can be generated from a 5 bladed

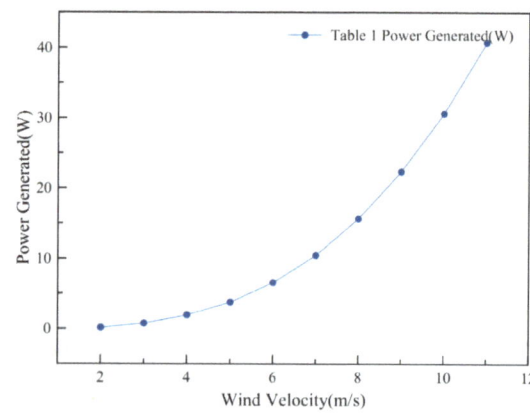

Fig.7 Graph of simulated power generated as a function of wind velocities varying from 2 to 12 m/s.

turbine with Tip Speed Ratio of 5, having blade efficiency of 0.5 and Blade length equal to 0.3 m at a wind velocity of 12 m/s. When we compared different values then keeping every other parameter same, the power generated was 114 W for wind velocity 11 m/s. However a decrease in the blade radius to 0.2 m, led to lower values of power generated (66.2 W) even if we increase the wind velocity to 12 m/s. This shows that blade radius should be of the order of 0.3 m. Hence it is concluded that a number of small turbines connected to a single axle with above mentioned specifications can be installed at Saket metro platform in order to generate a good amount of power.

ACKNOWLEDGMENT

Authors are thankful to University of Delhi for financial support. Authors thank Mr. S. A. Verma, the mentor of the project for his suggestions and guidance. Authors are also grateful to Delhi Metro Rail Corporation (DMRC) for their active cooperation in collaboration.

REFERENCES

[1]   How wind energy could fuel future power needs: M. Ganguly *et al.*; presented at the National Conference on "Striving & Thriving Towards Diffusion of Student-driven Research in Science and Technology for Inspired Learning" organized by Embedded Systems & Robotics Centre (ESRC) and Department of Electronics Maharaja

[8]

Agrasen College, University of Delhi held on 16-17 October, 2014. Proceedings ISBN No. 978-81-7273-958-4 pg 171.

[2]   "Feasibility Study of Conversion of Wind Energy to Electrical Energy at Delhi Metro Stations using Light Rotor Turbines", S. Patwal, *et al.*; Presented at Second National conference on "Student driven research for inspired learning in science and technology" organized by Embedded Systems and Robotics Centre and Department of Electronics, Maharaja Agrasen College, D.U. held on 16-17 October, 2015. Proceedings available at: http://www.ijsrd.com/articles/NCILP033.pdf. pg 131-136.

[3]   Wind Velocity Profiling at Delhi Metro Stations, P. Verma et al. Presented at National conference on "Advancements in Electronics and Computer Applications (NCAECA-2016)" organized by Department of Electronics and Computer Science, Shaheed Rajguru College of Applied Sciences, D.U. held on 4-5 February, 2016. Abstract book pg 69-73.

[4]   http://www.anylogic.com/features

[5]   http://m-selig.ae.illinois.edu/propid.html

[6]   http://windbench.net/models/sowfa/sowfa-les

[7]   https://www.warlock.com.au/tools/bladecalc.htmhttp://www.anylogic.com/features

Low Cost Eco-Friendly Solar Inverter

A Standalone Solar Power System For Households

Archana Rajput, Ayushi Chopra, Diksha Pandey, Pratima Kumari, Roopal, Smriti Srivastava, Vaishali Pathak (UG students)

Faculty Team: Dr Sneha Kabra, Ms Himani Dua, Ms Ritika Chopra
Shaheed Rajguru College of Applied Sciences For Women
(University of Delhi)

ABSTRACT: The demand of renewable, clean, highly efficient and stable power system is on rise due to various environmental factors. The presently available systems are either very expensive, or not efficient enough for large power operations, or are either not standalone. The objective of the proposed work is to develop an eco-friendly inverter, which would run completely on solar energy, which can be used to supply power to household appliances in the absence of electricity. The designed system can be used independently without any requirement of external power supply. The key challenge is to make the device at a low cost so that it is easily affordable by common man. The device once developed can also be used in rural areas where there is insufficient power supply.

KEYWORDS: Solar panel, Battery charge controller, DC to AC inverter, Low pass filter, 555 timer, ICs

I. INTRODUCTION

Solar energy is extremely sought after as it is economical, user-friendly, widely available, renewable, more cost effective, non-polluting and also easily accessible in remote areas. Due to various advantages, solar appliances have become very popular like solar refrigerators, solar geysers, solar cookers, solar calculators, solar street lights etc. Solar inverter is one of the solar energy powered device, which is yet to be explored for commercial and domestic applications.

Such an inverter would find applications in a lot of areas. For example, it can be used in rural areas where there is insufficient power supply. It can also be used for irrigation, water pumping in villages, water quality monitoring and environmental data monitoring. Solar inverters, in particular, would be very useful in case of natural calamities, when power grid gets damaged and there is no other alternative source of electricity.

In the proposed work, we plan to develop a low cost solar inverter, which can be used to operate all household appliances in the absence of electricity from the power grid. A normal inverter or a UPS has a battery to store energy from the grid when the supply is available and when the grid is cut off, it provides power from the charged battery. The circuitry of standalone solar inverter system would consist of a solar panel, battery, charge controller, inverter and the load.

The photovoltaic cell acts as a photosensitive diode that instantaneously converts light into electricity (DC). The energy generated is in raw direct current form. Which is converted into alternative current (AC) by using power inverters such that the resulting output AC is synchronized for being fed to electrical grid systems. The amount of power generated can be first tested on different loads and then it can be used to drive various appliances in households. To take the work further, an automatic solar tracker, which would rotate a solar photovoltaic panel or lens towards the sun, would also be included, since the position of the sun varies throughout the day.

Various projects and research papers [1][2][3][4] have been referred and the one's which are best suited to comply with our requirements (such as power capabilities, cost limitations and other parameters) have been selected, read and further modified to reach the required output.

Circuit Schematic:

CHARGE CONTROLLER CIRCUIT (Fig.1): *converts DC output of PV panels into a clean AC current for AC appliances.*

In Figure 2 is shown a transformer and LM555 IC based inverter circuit with configuration capable of giving a 50 Hz 220V output. Where, V1 is a 12V DC input voltage supply which can be from any renewable energy generator such as a solar panel, resistors R1, R2 and capacitor C1 are acting as a time constant, the transistor Q1 is NPN and

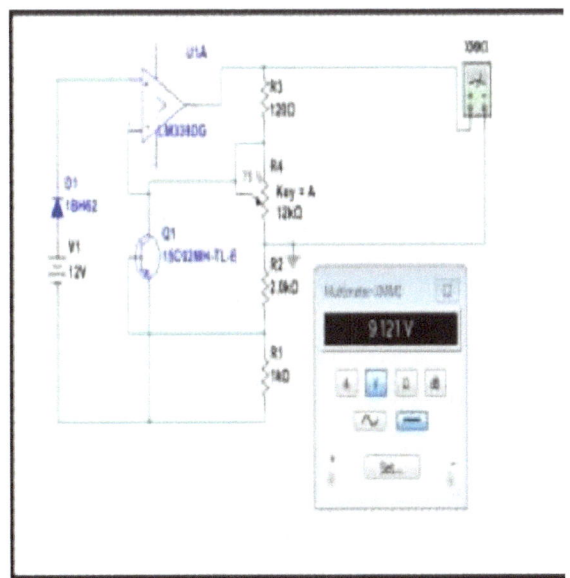

Fig. 1: Charge controller circuit

while transistor Q2 is PNP, both make up a pair of switching devices to generate

AC cycle, while inductor L1 and capacitor C4 are acting as a low pass filter to convert square wave output from the transistor-pair into sinusoidal wave

Charge controller circuit regulates the voltage and current going to the battery which is coming from the PV panels. It also prevents battery overcharging and thus prolongs the battery life.

The given input in this circuit is 12V which is to be obtained from the solar panel. The output obtained is 9.121V

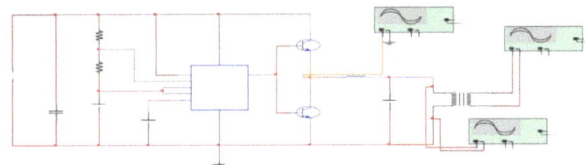

Fig. 2 : LM555 IC based inverter circuit

INVERTER CIRCUIT (Fig. 2): *Inverter circuit*

Wave Shaping: The output from NPN and PNP transistor pair is in square wave which also contains harmonics and wave form disorder properties as shown in Figure 3.After amplification by the

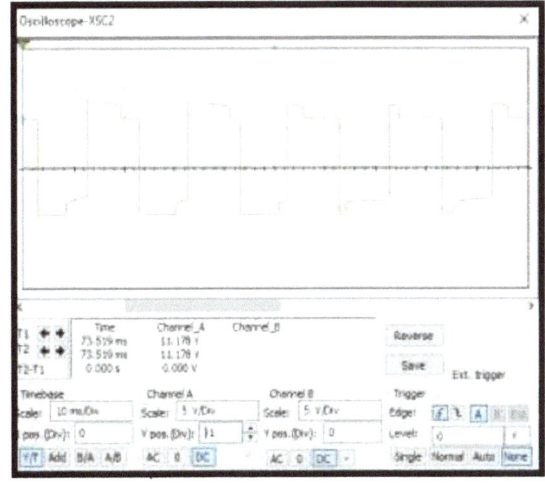

Fig. 3: Output waveform at transistor terminal

transistors, the proposed low pass filter

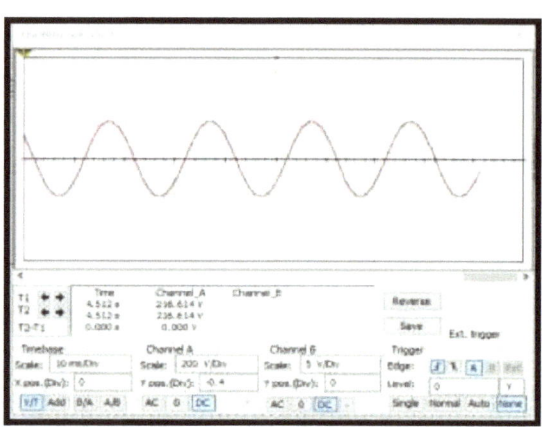

Fig. 4: Inverter 220v 50Hz ac output

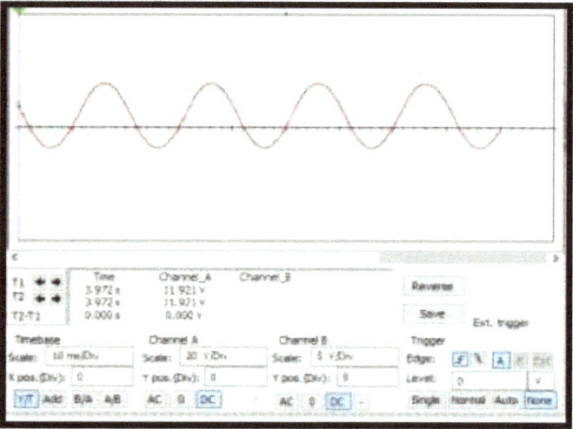

Fig. 5: Output waveform after low pass filter

configuration for 50Hz frequency the harmonics are reduced and the output waveform is in proper sinusoidal shape as shown in figure 5[4].

II. RESULT AND DISCUSSIONS

The output from the low pass filter which is coming as 3v (practically) has to be further amplified to 12v, for which we will add a voltage amplifier whose output is then passed through to a transformer to step it up to 220 V AC. Figure 4 shows the results for R1=10k and R2= 139k with frequency of 50Hz with no load attached (simulation based result).

III. CONCLUSION

This paper includes a LM555 timer based inverter circuit to generate 50Hz AC output and a charge controller circuit. Also the low pass filter design is to transform distorted square wave into pure sinusoidal wave form. All the waveforms shown are simulation based results (Multisim14) and provide a final 220V proper shaped output.

ACKNOWLEDGEMENT

This work is supported by Research Grant for Innovation project (project code: SRCA-316) by Delhi University. The authors would like to thank University of Delhi for giving financial assistance to carry out the work. The authors are also thankful to Professor Mridula Gupta, Department of Electronic Science,South Campus,University of Delhi.

REFERENCES

[1] Rodriguez, J., S. Bernet, B. Wu, J.O. Pontt and S.

[2] Kouro, 2007. "Multilevel voltage-source-converter topologies for industrial medium-voltage drives," IEEE Trans. Ind. Electron., 54(6): 2930-2945.

[3] Sarvi, M. and M. Keshmiri, 2013. "A Fuzzy-PD Controller to Improve the Performance of HVDC

[4] System" World Applied Sciences Journal, 22(9):1210-1219.

[5] RohollahAbdollahi,2013."ANovel Delta/Hexagon-Connected Transformer-Based 72-Pulse AC-DC Converter for Power Quality

[6] Improvement", World Applied Sciences Journal,23(3): 390-401.

[7] World Appl. Sci. J., 30 (Innovation Challenges in Multidiciplinary Research & Practice): 141-143, 2014.

[8] http://mselig.ae.illinois.edu/propid.html

Index of Authors

Sponsors

DBT

College

Star Scheme

The organizing committee of RTIE 2016 is grateful to our generous sponsors without their support it is not possible to have a conference of this stature.